PHILOMATH

PHILOMATH

The Geometric Unification of Science & Art through Number

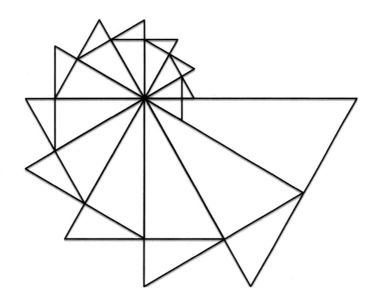

Philomath

The Geometric Unification of Science & Art through Number

Cover design by Will Weyer

ISBN-13: 9798736939466

First Edition, April 2021

Printed by KDP, Amazon.

As Above...

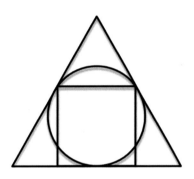

CONTENTS

About the Authors

Robert E. Grant

Mr. Robert Grant holds a BA from Brigham Young University and an MBA from Thunderbird, The American Graduate School of International Management. He was formerly CEO and President of Bausch and Lomb Surgical. Robert is the Founder, Chairman, and Managing Partner of Strathspey Crown LLC, a growth equity holding company with a broad portfolio of company and asset holdings spanning healthcare, clean energy, social media, and financial technology. In addition, he is the Founder, Chairman, and CEO of Crown Sterling Limited LLC, a next-generation cryptography company based on novel discoveries in geometry and mathematics. He has multiple publications in mathematics and physics related to his discoveries of quasi-prime numbers (a new classification for prime numbers) and a unification wave-based theory connecting and correlating fundamental mathematical constants. He is also an accomplished artist and musician.

Talal Ghannam, Ph.D.

Dr. Talal Ghannam is a physicist and number enthusiast. He got his bachelor's in mechanical engineering from Aleppo University, Syria. He completed his master's and Ph.D. in theoretical physics at Western Michigan University. His research concentrated on lasers, quantum optics, nanotechnology, and plasmonic sensors. His numeric research revolves mainly around digital root math, exploring how it correlates with natural sciences and its ability to explain many of the most fundamental physical principles. He wrote unique books on this subject, with the latest titled *GeoNumeronomy*, where the simple nine numbers of the digital root operation and the geometric patterns they make are used to explain the universe including the ether and the fourth dimension, the subatomic particles, the fundamental forces, and the DNA. He is also a data scientist, independent historian, and comics creator.

With special thanks to our wives, children, and parents for allowing us to think outside the box and for teaching us a love for geometry and mathematics. At times, our research required late nights and many weekends, at significant sacrifice to family and friends. We are grateful that our loved ones afforded us this opportunity.

We would also like to acknowledge with special thanks, Will Weyer, Alan Green, Adam Apollo, Jamie Janover, Jonathan Leaf, and John Wsol for their many contributions to our research.

NUMBERS

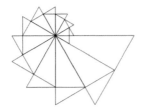

KEYS TO THE DIVINE ENCRYPTION

Since the dawn of time, humans have counted.

They counted their fingers and household, animals and cattle, days and nights, months and years, and so on. Whether consciously or subconsciously, the aspect of numbers has always been wired into their brains. Initially, numbers were expressed in limited terms, such as one, two, and many. Since then, however, they evolved greatly, stretching out toward infinity, and at the same time were given all sorts of shapes and forms to be easily represented and manipulated. More importantly, though, they were given meanings and symbolism.

The relationships humans found between the world around them and their invented tool of numbers were plenty and fundamental. From their own body parts to the plants and their seeds and yields to the cycles of the moon and the periods of the sun and planets, all awakened inside them a sense of a connection and meaning behind the universe. Moreover, they realized that geometry and numbers are not separate entities; they are strongly connected and interlinked. Can we think of a pentagonal flower without invoking number 5 in the back of our heads? The Fibonacci sequence is purely numeric in its essence; however, once its numbers are mapped into dimensions of squares, a beautiful spiral found abundantly in nature can be traced within. Fundamental constants like π and the golden ratio Φ are infinite numbers that govern geometrical forms and their expansions. In fact, whenever there is a shape or form that resonates with us humans, its dimensions and proportions would almost always be based on magical numbers or fundamental constants.

Believing that numbers and geometry emanate from the same source, there must be a reason why certain number/shape combinations were chosen for this creature while others were chosen for that. Thus, humans saw in nature a puzzle, a divine encryption whose deciphering keys are scattered all around, hidden within the complex layers of the world they perceive through their senses.

As we receive geometrical patterns through our eyes, one way they are perceived within our brains is through numbers, e.g., 3 for a triangle, 4 for a square, etc. Those shapes that cannot be described as such will mostly be considered random or non-uniform. Even for a 3-dimensional sphere, if the two perpendicular diameters of its 2-dimensional mental-projection are not equal -do not have a ratio of unity- then it is not a circle and, consequently, not a true sphere. This could very well be how our brains work; they transform everything into numbers and ratios subconsciously, and based on these values, they decide on proportions, distances, uniformity, etc. And if these ratios turned out to be based on magical numbers, the experience would transcend our brains and delve deep into our souls.

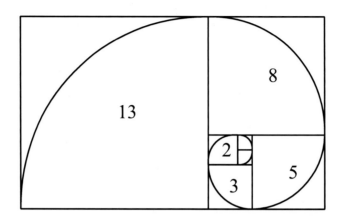

Figure 1: A perfect spiral made from adjacent squares having dimensions based on the Fibonacci sequence.

Hence, numbers are one conduit by which the sight, mind, and soul are interconnected. And through this sense, man found the second layer of the divine encryption, geometry, where the first layer is but numbers themselves.

Hearing is the next most important sense humans possess (some may argue it is the first most important). There are several properties that quantify the phenomenon of sound. There is the frequency, pitch, amplitude, etc., all of which are described through numbers. But in order for a combination of sound tones to feel pleasant and harmonious to our ears, to become music, the numbers that describe them must have specific values and ratios between each other. This is what Pythagoras is believed to had discovered more than two millennia ago, that particular ratios of notes produce enjoyable experiences, while others do not. Therefore, even though the hearing experience, in general, is not based entirely on numbers, what is pleasant to the human ear and soul depends heavily on them. In this sense, geometry and sound are similar, as for them to be received well by our brains and resonate with our souls, they must correspond to specific numbers and ratios. (And as we to discover later, geometry is very much interconnected with sound, especially music.)

Thus, man discovered the third layer of the divine encryption: sound and music.

Those numbers humans repeatedly observed in nature, whether in form or sound, held a special place within their conscious and subconscious minds. They were deemed *Magical*, emanating from the divine intellect to manifest as wonderful creations in our earthly domain. From this belief, scientists, philosophers, and artists embarked on a mission to incorporate these magical numbers within their works, whether a book, a painting, or even a musical piece. For example, many authors arranged their books' chapters based on specific numeric sequences, like that of Fibonacci. Sometimes, they would put hints or references to these numbers inside their texts by repeating certain words or capitalizing specific letters, etc. Some musicians are believed to have composed their pieces based on specific numeric ratios and mathematical constants, e.g., the golden ratio.

But none exploited the power of numbers and their meaning better than painters and architects. It is in these visual domains that numbers found fertile grounds where they could be deeply seeded, whether in a painting or a building, to endow it with energy, make it harmonious to the gazer, and of course, hide messages and clues that may lie dormant for hundreds or even thousands of years before the sharp eyes of the seeker awaken them.

Ancient buildings and temples have always been living testimonies for the number-obsessed minds behind their creation. Structures like the Great Pyramid of Giza or the Parthenon of Greece are filled with constants of nature, from π to e to Φ, to mention a few. To embed these numbers in the design and dimensions of their buildings, the architect and masons required excellent mathematical knowledge and engineering skills. Therefore, magical numbers were not only the result of keen observation and contemplation, but they also motivated the research for more advanced mathematics, engineering, and science.

The final layer of the divine encryption came from natural sciences, where many fundamental numbers, labeled constants, were found to control life and the whole universe, such as the fine structure constant α, which controls the interaction of light and matter; the Euler constant e, which controls radioactive decay as well as many other natural phenomena; the Planck constant $\hbar$, and the Coulomb constant C, to mention a few. The quantization of the atomic domain opened the door to a fascinating and mysterious world where numbers ruled unchallenged. Everything in the atom is governed by specific Quantum Numbers that dictate how its internal structure, which includes the electrons, protons, and neutrons, interacts and behaves. Some of these numbers are so special that they were labeled magical by the scientists, such as the magical numbers of the nuclei [2, 10, 18, 36, 54, and 86]. And as we will discover, many of these numbers were already known and admired for millennia, observed on the macroscopic level well before the microcosmic one.

With geometry, sound, and science, all united through numbers, the layers of the divine encryption were complete, and it was the right moment for the keys to be discovered and the encryption to be deciphered. Unfortunately, this complete knowledge came at a time when numbers were stripped of their meaning and symbology, when they became mere tools and digits of calculation. They were deprived of their true powers, their ability to bring all the senses together and unify all knowledge into one holistic matrix.

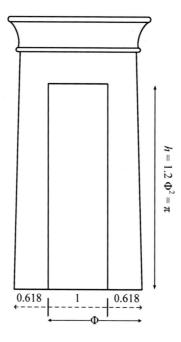

Figure 2: The golden section Φ and the π constant embedded within the dimensions of the gates of Egyptian temples.

In correspondence with this sterile attitude, schools' and universities' curriculums were designed to intentionally omit the true purpose of numbers, and the various scientific fields were torn apart, specialized into narrower and narrower topics, which deprived the knowledge seeker of experiencing the real aim of science and exploiting the full benefits of its holistic approach. Fragmentation became the spirit of the age.

Motivated by their love and appreciation for numbers and the meaning they add to our understanding of the universe, the authors of this book embarked on a journey of learning that took them across space and time, where amazing discoveries were made and uncharted territories were mapped. They joined forces to bring this knowledge to the general public and make it accessible to all. They believe that now, more than ever, humanity is in dire need of letting go of its fragmented mentality and embracing the natural holistic view of the universe. From this context, the authors strived to make this book accessible to the young and the old, the novice and the expert alike. They used as simple language as possible, trying to avoid the complex and often incomprehensible languages and terminology most advanced scientific books are plagued with. In fact, the authors hold a strong conviction that true knowledge can only be explained by the simplest of words. In addition, plenty of illustrations were provided throughout the book to better explain the ideas and to make sure abstract concepts are always coupled with visual inputs to create a permanent imprint of the absorbed information.

The book is organized into four main parts corresponding to the four layers of encryption mentioned above. The first part is about the numbers and the mathematical ideas and theories upon which the rest of the book stand. This is an essential part, and as we do not recommend skipping any part or chapter in this book, this particular one should be thoroughly understood and definitely not skipped. The second part is about geometry in general and its fundamental relationship with numbers. It also concentrates on the topic of sacred geometry, which is the geometry of shapes believed to have the most significant presence and effect on our lives, and especially on their unique numeric properties. The third part is about sound and music and the strong bonds they have with numbers and geometry. The fourth part covers natural science, mainly the physics of the atomic domain. In addition, it will be shown how the main forces of nature emanate from and are ruled by the geo-numeric concepts explained in the previous parts.

A fifth part explores the extent to which the knowledge of numbers and geometry has transcended the boundaries of space and time by tracing their presence in the works of one of the most visionaries of the renaissance age, Leonardi Da Vinci.

This book is a wonderful journey through which the reader will confront novel ideas and uncharted scientific territories that will challenge his long-held beliefs and expand his mind toward limits he never thought possible. In the process, he will be rewarded with a completely new vision and understanding, not only of numbers and science but also of his own unique existence.

PART I

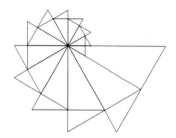

MATHEMATICS AND NUMBERS

"All is number."
-Pythagoras of Samos

We start the journey with the numeric and mathematical foundations upon which this book stands. As mentioned in the introduction, the theories and information provided in this chapter are essential for the proper understanding of the rest of the book. They are mostly already known and simple scientific concepts; however, we dwell on them, expanding them into new territories uncharted by the orthodox scientific cabal.

For example, the digital root of numbers is an already known branch of number theory that has been studied for millennia. In this book, however, we expand it into areas that are way ahead of any conventional view on the subject, proving it to be a far more important discipline than previously thought. And if digital root math, in essence, is a well-known field, a concept like *wave-particle duality* has not been considered a property that could be applied to simple numbers; it has been applied to numerous physical phenomena, but not to pure digits. In this part, we show how this bizarre and quintessential physical phenomenon does, in fact, apply to, if not emanates from, the wave-particle duality of numbers themselves. We will also investigate prime numbers in depth, numerically, and, more importantly, geometrically, which is a must if we want to fully understand these unique numbers. And while the reciprocal of numbers is an operation that has been deemed trivial by most mathematicians, here, however, we show it to be one very important concept that definitely deserves deeper study and investigation.

The above topics, and more, will be explained here and used extensively throughout the book. They will serve as a blueprint through which we will navigate the reader into the uncharted territories of math and numbers that are so essential to our correct understanding not only of mathematics but most physical sciences as well.

The Digital Root

*"Mathematics is the queen of the sciences
and number theory is the queen of mathematics."*

-Carl Friedrich Gauss

Math and Properties

We begin our math and numbers discussion by explaining the fundamentals of the digital root as it will be used extensively throughout the book.

The *digital root*, also known as *modular math*, is one important branch of number theory. It has many applications, especially in the field of digital security and encryption. Modular math can work with any numerical base, e.g., base-10, base-12, base-60, etc. In this book, however, we concentrate mainly on base-10 (which is actually base-9, as explained below.)

In simple words, what the digital root operation does is to reduce a number down to a single digit. This is done by continuously summing the number's individual digits. For example, the digital root of 5432 is calculated as: $5 + 4 + 3 + 2 = 14 = 1 + 4 = 5$. The possible outcome of this operation can be any number from 1 to 9 only. (This is why we suggested it is more of a base-9 rather than base-10.)

Under this operation, we can think of numbers occupying a wheel of 9 spokes and segments (better known as *moduli*, plural form of *modulus*) where numbers from 1 to infinity are distributed around the nine moduli. In this configuration, each modulus houses only those numbers with the same digital root, as shown in the figure below.

Before we proceed, here are a few notes on the digital root terminology we will be using throughout the book: the letter D indicates taking the digital root of the number; e.g., $D(13)$ means to take the digital root of number 13. The term *D-sum* refers to the digital root of the sum of a group of numbers. So,

D-sum(13, 14, 15) = *D*(13+14+15) = *D*(42) = 6. And *D-space* refers to the space of the digital root operation, where numbers range from 1 to 9 only.

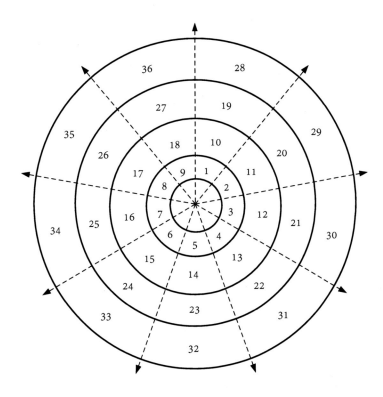

Figure 3: The digital root wheel. Those numbers that have the same digital root occupy the same modulus.

Below we explain the most basic algebraic operations within the logic of digital root math.

<u>Addition</u>

If a + b + c + … = d, then $D[D(a) + D(b) + D(c) + …] = D(d)$.

Example: 3830.64 + 9457.385 = 13288.025.

$D(3830.64) = D(3+8+3+0+6+4) = D(24) = 2+4 = 6$, $D(9457.385) = D(41) = 5$, and $D(13288.025) = 2$. Therefore $D(6 + 5) = D(11) = 2 = D(13288.025)$.

It is obvious that the digital root of any irrational number is undefined, like in the case of π, for example, where $D(\pi) = D(3.14159265…)$ And because π is infinite, its digital root is something that cannot be calculated.

4

Multiplication

If $a \times b \times c \times \ldots = d$, then $D[D(a) \times D(b) \times D(c) \times \ldots] = D(d)$.

Example: $35.86 \times 95 = 3406.7$ and $D(35.86) = 4$. $D(95) = 5$, $D(3406.7) = 2$, and $D(4 \times 5) = D(20) = 2 = D(3406.7)$.

Subtraction

Rule: If $a - b - c - \ldots = d$, then $D[D(a) - D(b) - D(c) - \ldots] = D(d)$.

Division

The division operation works fine for most numbers but not always, e.g., when we divide by 7. (Generally, in modular math, the division operation depends on the greatest common divisor between the numbers involved and the numeric base. However, we will keep it simple here as it is not much needed in the context of the book.)

Example: $3456/5 = 691.2$ and $D(3456) = 9$ and $D(9/5) = D(1.8) = 9$.

Dividing by 7 is a special case, where the result is always a repetitive sequence of the numbers: [1, 4, 2, 8, 5, 7]. These six digits [1, 4, 2, 8, 5, 7] are called the *septenary numbers,* and we will be encountering them a lot throughout the book.

Number 9: The Zero of the *D*-space

Even though number 9 is the last and largest number of the *D*-space, it behaves in this finite space as number 0 does in the regular infinite numerical space. For example, adding 9 to any number will not change its digital root, just like adding 0 to any number has no effect on that number either, e.g., $D(276) = 6$, and $D(9+276) = D(285) = D(2+8+5) = D(15) = 6$. Furthermore, the digital root of multiplying any number with the number 9 will always be 9, just like multiplying 0 with any number, and no matter how large it is, reduces it to 0, e.g., $D(241 \times 9) = D(2169) = 9$.

This property of number 9, being the largest and smallest at the same time, is unique to the *D*-space and has some interesting implications, as we discover later on.

Throughout the ages, the number 9 used to symbolize completion; it is the number that ends cycles of times, bringing one age to an end before a new era begins, as in the procession of the equinox, which is the movement of the zodiacal signs behind the rising sun. Every 72 years ($D = 9$), the zodiacs move one degree from their previous year's position. For a full cycle, it takes around 25920 years (360×72), and $D(25920) = 9$. Currently, we are entering or already in the age of Aquarius.

Berossus, the Babylonian priest who lived around the 3rd century BC, listed the ten Sumerian antediluvian kings, showing them all to have ruled for periods whose digital roots are 9, as shown in the table below.

King's Name	Ruling Period
Babylone	36.000 years
Babylone	10.800 years
Pautibiblon	46.800 years
Pautibiblon	43.200 years
Pautibiblon	64.800 years
Pautibiblon	36.000 years
Pautibiblon	64.800 years
Larak	36.000 years
Larak	28.800 years
Shuruppak	64.800 years

Table 1: Berossus' list of the antediluvian Sumerian kings and their ruling periods.

In Greek mythology, Zeus inflicted a great flood on Earth to wipe out humanity, which lasted 9 days. In the Bible, Noah is stated as the 9th patriarch before the great deluge engulfed Earth. Similarly, all the Mayan time periods have digital roots of 9, as well as the Hindus Yugas: Satya Yuga = 1728000 years, Treta Yuga = 1296000 years, Dwarpa Yuga = 864000 years, and Kali Yuga = 432000 years.

Whether these periods are real or not, doesn't matter; what really matters is the role number 9 plays in completing all these cycles, the one numeric property believed by all civilizations around the world. There is something fundamental about this number that is deeply rooted in humanity's subconscious, and this book may shine a light on the reason.

The Basic *D*-Circle

As we saw in the previous chapter, numbers in the *D*-space do not keep increasing towards infinity; instead, they go around from 1 to 9 back to 1 in an infinite loop. In doing so, they form circles (or spirals) like the one shown below, which exhibit many interesting properties.

For example, numbers [3, 6, 9] divide the circle into three equal segments where the *D*-sum of the numbers in each segment is also 3, 6, and 9: 1+2 = 3 , 4+5 = 9 and $D(8+7) = D(15) = 6$. Moreover, the *D*-sum of every two numbers facing each other across every diameter is the same. For example, the *D*-sum of each two numbers facing across the diameter that starts with 9 (which we call the 9-symmetry line) is equal to 9: 1+8, 2+7, 3+6, 4+5 (as shown in the same figure); across the 3-symmetry line is 6: 2+4, 1+5, 9+6, 8+7; across the 6-symmetry line is 3; across the 1-symmetry line is 2, and so on. The rule is: numbers across any symmetry line will add up to double the number associated with that line. Moreover, numbers [3, 6, 9] divide the circle into three segments, with the *D*-sum in each segment being equal to [3, 6, 9] as well.

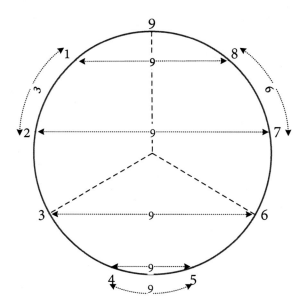

Figure 4: The nine digits of the *D*-space repeats in a circular or spiraling fashion. Numbers [3, 6, 9] divide the circle into three parts where the *D*-sum of the numbers in each is also 3, 6, and 9.

If we start with number 1 on the basic *D*-circle, drawing a line to number 2, doubling our steps on the way, we go to number 4, then 8. Doubling again gives us 16, with a digital root of 7, so we connect to 7. Doubling more gives 2×16 = 32 with a *D* of 5, then 64 with a *D* of 1, which brings us back to our starting point. Doubling more takes us along the same path again, ad infinitum. In the process, we trace a V-shape (or the infinity symbol ∞) made of the numbers [1, 2, 4, 5, 7, 8], the same septenary numbers we saw in the previous section. Throughout the doubling process, the group of [3, 6, 9] becomes isolated from the rest. The doubling of this group is also different as 3 doubles to 6, which doubles to 3 back, and so on. Number 9 doubles into itself only.

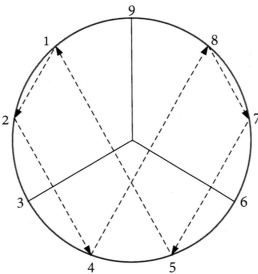

Figure 5: Within the *D*-space, the doubling process starts from number 1, going to numbers 2, 4, 8, 7, 5, and back to number 1 again.

Arranging the six digits of the doubling operation around a circle in their exact order (2, 4, 8, 7, 5, 1) leaves even numbers on one side of the circle and odd ones on the other. Odd and even represent the numeric aspects of the opposing forces of the universe, similar to the Yin-Yang concept of the Tao philosophy, which symbolizes the duality that exists in the universe: water and fire, wet and dry, dark and light, male and female, etc. In order to maintain the harmony and balance of the whole universe, these opposing forces intertwine and overlap with each other in a circular or wave-like pattern, as in the Yin-Yang symbol, shown below.

The two small white and black circles represent each force's existence within its opposite, nullifying the absolute purity of each force while elaborating on the need each one has for the other to complete itself.

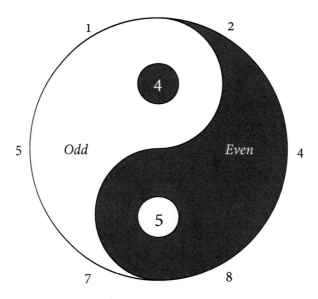

Figure 6: The Yin-Yang symbol of the opposing aspects of nature, expressed via a perfect numerical language.

We can rephrase the above logic numerically by noticing that when we add the even numbers together (2+4+8), we get 14, which has a digital root of 5, an odd number. Doing the same for the odd ones *D* (1+5+7) results in number 4, which is even. This is why these numbers are positioned at the centers of the small circles. Furthermore, adding the odd and even *D*-sums to each other (4+5) results in number 9, a number that brings completion. Adding up the diagonal numbers generates number 9 as well: 1+8, 2+7, and 4+5. Thus these opposing aspects need each other to become complete in a conditional unity. (This specific completion property will prove essential for our understanding of the universe, especially when we discuss the atomic domain.)

The above numeric correspondence is just one example of many that illustrate how simple numbers and their digital roots can verify ancient wisdom. Everything in the universe, from forces to matter to physical laws, has its own numeric correspondence. Numbers are the universal language; we only need to learn how to read it.

Prime Numbers

"Mathematicians have tried in vain to this day to discover some order in the sequence of prime numbers, and we have reason to believe that it is a mystery into which the human mind will never penetrate."

-Leonhard Euler

Properties and Distribution

As we saw earlier, numbers can be either odd or even. But this is not their only categorization; they can also be prime or composite, integers or floats, rational or irrational (or transcendental), etc. Nevertheless, primeness is one of, if not the most important of these categories.

By definition, a *prime number* is exactly divisible (with no remainder) by number 1 and itself only. Everything else is considered *composite*. And while prime numbers are always odd (except for number 2), composite numbers can be either odd or even. The first few prime numbers are: [2, 3, 5, 7, 11, 13, 17, 19, 23, 29, 31, 37, 41, 43, 47, 53, 59, 61, 67, 71, 73, 79, 83, 89, 97, 101, 103, 107, 109, 113, 127, 131, 137, 139, 149, 151, 157, 163, 167, 173, 179, ...]

The number 1 is not considered a prime by most mathematicians. This is because it is much easier to prove many of their mathematical properties if it is excluded. (Hence, number 1 can be considered a category by itself, neither prime nor composite.)

As is evident from this sequence, the frequency of prime numbers gets smaller and smaller as we continue toward bigger numbers. Still, no exact mathematical formula can predict their occurrences or generate them all. Some formulas can generate a few of them, but eventually, they all fail at some point. Thus, in addition to lacking factors, they also lack a mathematical pattern. This unique quality, among many others, makes prime numbers ideal for encryption and security purposes, e.g., in credit card information and secure internet surfing. Basically, this is because multiplying a couple of very large prime numbers to create an even larger number (called a *semiprime*) is an easy and fast operation. However, figuring out which prime factors are implemented in producing the semiprime is a very hard and lengthy process.

Even if no specific mathematical expression can generate all prime numbers, this doesn't mean their distribution is completely random. As we just mentioned, besides number 2, all prime numbers are odd because any even number can be exactly divisible by 2, and therefore, it is no longer a prime. Also, it is well known that the last digit of any prime number greater than 5 can only be 1, 3, 7, or 9. Numbers 2, 4, 6, and 8 are excluded to ensure that the number is odd. The number 5 is also excluded because any number ending with 5 will be divisible by 5.

Mathematically, prime numbers always come in the form of $6k \pm 1$, where k is any integer from 1 to infinity. Of course, not all k values produce prime numbers; otherwise, this formula would have been considered a perfect generator of these numbers. From this formula, it appears number 6 is a pivot or a center number around which prime numbers emerge. This specific role of number 6 is so fundamental that it is at the core of some of this book's most important theories.

Even though the above formula is not a perfect prime-generator, it can still be used to prove other prime-related properties. One such interesting property is that for any prime number p, its square is always a multiple of 24, plus 1 ($p^2 = f \times 24 + 1$). This can be proven as follows: the factor k in $6k \pm 1$ can be either even or odd, so we write it as $2m$ for even and $2m+1$ for odd. Next, we substitute these values in the main equation and raise it to the power of 2, which results in four forms for p^2: $(12m+1)^2$, $(12m+7)^2$, $(12m-1)^2$, and $(12m+5)^2$. When the squared terms are expanded, we get four terms that involve the number 24 multiplied by some factor of m, plus 1, such as $(6m^2+m) \times 24 + 1$ and so on for the other three. If we combine $(6m^2+m)$ into a single factor f, we retrieve the desired equation: $p^2 = f \times 24 + 1$.

Another prime-property is that in the D-space, with the exception of number 3, the digital root of any known prime number is anything but 3, 6, or 9. To prove this claim, we start with the fact that $p^2 = f \times 24 + 1$. Now, if p has a digital root of 3, 6, or 9, then $D(p^2) = 9$, always. On the right side of the equality, $D(24) = 6$ and $D(f \times 6) = [3, 6, 9]$ always. (This can be checked by looking at the digital root of the multiplication table, also known as the Vedic Square, shown in the table below.) By adding 1 to the right side of the above equality, we have [3 or 6 or 9] + 1 = [4, 7, 1], always. Therefore, we can never get the number 9 on both sides of the equation, and consequently, no prime number can have a digital root of 3, 6, or 9.

Thus, if we want to know whether a number is prime or not, we first make sure it is even, and if so, check its digital root. If it is either 3, 6, or 9, then the number is definitely not a prime. Otherwise, there is a possibility of primeness.

×	1	2	3	4	5	6	7	8	9
1	1	2	3	4	5	6	7	8	9
2	2	4	6	8	1	3	5	7	9
3	3	6	9	3	6	9	3	6	9
4	4	8	3	7	2	6	1	5	9
5	5	1	6	2	7	3	8	4	9
6	6	3	9	6	3	9	6	3	9
7	7	5	3	1	8	6	4	2	9
8	8	7	6	5	4	3	2	1	9
9	9	9	9	9	9	9	9	9	9

Table 2: The digital roots of the multiplication table for numbers from 1 to 9, also known as the Vedic Square. Notice that multiplying any number by 3, 6, or 9 will generate a digital root of 3, 6, and 9.

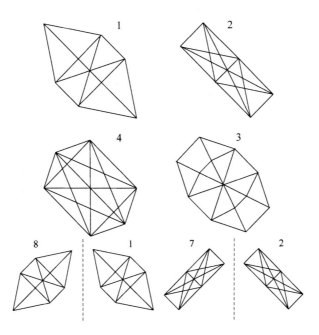

Figure 7: The Vedic Square was used to check calculations' validity as well as to generate interesting geometric patterns by highlighting specific numbers and then connecting the cells containing these numbers. Shapes that correspond to numbers adding up to 9, as [2, 7], [1, 8], etc., are mirror images of each other. Tracing number 9 generates a square, which is its own mirror image.

There are many methods to generate prime numbers. One famous method is the Sieve of Eratosthenes, attributed to the Greek mathematician Eratosthenes of Cyrene (276 - 195 BC), which offers an easy but lengthy trick to find all prime numbers that are less than a certain number. For example, to find the prime numbers within the range from 1 to 42, we first write these numbers in a grid, as shown below.

1	2	3	4	**5**	6
7	8	9	10	11	12
13	14	15	16	17	18
19	20	21	22	23	24
25	26	27	28	29	30
31	32	33	34	35	36
37	38	39	40	41	42

Table 3: The Sieve of Eratosthenes' grid for picking prime numbers in the range from 1 to 42.

Next, we take out all numbers that are multiples of 2 (besides 2, all primes are odd). This will exclude [4, 6, 8, 10, 12, 14, 16, 18, ..., 42] from the grid. Then we take out all the numbers that are multiples of the number that comes after 2, which is 3, so we take out: [9, 15, 21, 27, 33, 13]. This is because, after number 3, no other number should be divisible by 3, or it will not be prime. We keep doing the same for the next numbers of 5, 6, and 7. The first multiple of 7 after 35 is 49, which is bigger than 42, and consequently does not belong to the grid. Hence, the remaining numbers are the prime numbers that lie within the range 1 to 42: [2, 3, 5, 7, 11, 13, 17, 19, 23, 29, 31, 37, 41] (excluding the number 1 of course).

Notice from the above grid how all prime numbers fall within the 1st and 5th columns only. This is because those falling in the 2nd and 4th columns are even, and those falling in the 3rd and 6th columns all have digital roots of [3, 6, 9]. As we will see shortly, when numbers are distributed in columns or moduli whose numbering is a multiple of 6, prime numbers will occupy specific moduli that follow unique geometrical patterns.

Another prime pattern that is geometric in its essence was found by Stanislaw Ulam (1909-1984 AD), a Polish mathematician who discovered that by plotting numbers in a spiraling sequence, prime numbers tend to line themselves on diagonal lines, as shown below.

In recent years, scientists and mathematicians have found more complex orders hidden within the ostensibly random distribution of prime numbers. One such order has been discovered by two mathematicians from Stanford University, Kannan Soundararajan and Robert Lemke Oliver. As we have seen, one known property of prime numbers is that they must end with either 1, 3, 7, or 9. So, if prime numbers occur in a purely random fashion, it shouldn't matter what the last digit of the previous prime is to the next one; each one of the four possibilities should have an equal 25% chance of appearing at the end of the next prime number. However, this is not what they have found. After looking for the first 400 billion primes, they found that prime numbers tend to avoid having the same last digit as their immediate prime predecessor.

As Dr. Oliver put it, they behave as if they "really hate to repeat themselves." Moreover, they found that primes ending with 3 tend to be followed by those ending with 9, instead of 1 and 7. (This behavior may be explained through the digital root operation, as will be explained later.)

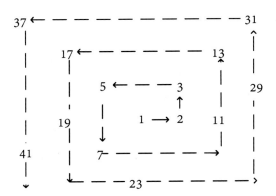

Figure 8: Ulam spiral. When prime numbers are plotted in a spiraling form, prime numbers fall on diagonal lines.

Another recent and surprising prime-order was discovered by the theoretical chemist and Princeton professor Salvatore Torquato. Prof. Torquato had the idea of looking at what diffraction patterns prime numbers would produce if they were to be modeled as atom-like particles. He observed that not only do prime numbers create a quasicrystal-like interference pattern, but the pattern itself is also a self-symmetry fractal that has never been observed before, which he labeled *effectively limit-periodic*. Prof. Torquato stated : "There is much more order in prime numbers than ever previously discovered."

It is evident from the above that prime numbers are not really pattern-less. On the contrary, there are so many conditions and rules that dictate their distribution, especially in regard to number 6. And it is in the geo-numeric space, when geometry and numbers are combined, that their hidden patterns become visible, as we explore next.

The 24-Wheel and Quasi Primes

In his book *God's Secret Formula*, Dr. Peter Plichta showed that by distributing numbers around concentric circles, 24 in each, prime numbers will position themselves along specific diameters or moduli, making a shape that looks like a Maltese cross or a plus sign. The pattern is infinite and can be proven as follows: any number on the 2nd modulus can be written in the form of $2 + 24h = 2(1+12h)$, where h is an integer from 0 to infinity, and therefore it will always be divisible by 2. Consequently, it cannot be a prime. The same logic applies to moduli 3, 4, 6, 8, 9, 10, 12, 14, 15, 16, 18, 20, 21, and 22, leaving the remaining eight moduli, the *prime moduli*, as the only placeholders for prime numbers. Those primes that reside on the same arm of the plus sign and differ by two are called *Twin Primes*, while those that reside on adjacent arms and differ by four are called *Cousin Primes*.

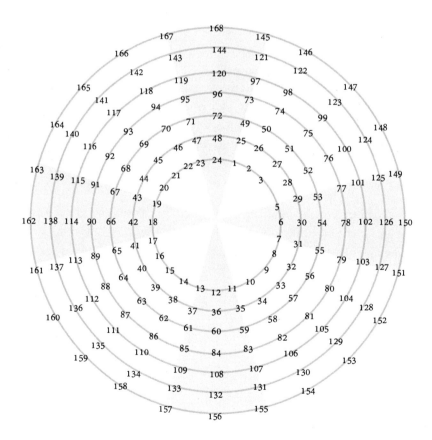

Figure 9: Prime numbers distribution around a 24-moduli circle will occupy moduli that form the shape of a Maltese cross.

In general, prime numbers will form shapes having a symmetry equal to the number of moduli, divided by six. So, for the above example, prime numbers form a cross, and $4 = 24/6$. When numbers are distributed around a 30-moduli wheel, prime numbers will form a pentagon (5) as $5 = 30/6$, and so on. This is, of course, a consequence of the property mentioned earlier, where any prime number comes in the form of $6k \pm 1$.

The 24-based distribution appears to possess unique properties when compared to others. For example, all prime-squared numbers reside within the 1st module only. This is not the case for other distributions, even if they are based on number 6, like 18 or 30, etc. This behavior emanates from the fact that squaring any prime number will always result in a multiple of 24, plus 1.

Those numbers residing along the prime moduli (the 1st, 5th, 7th, 11th, 13th, 17th, 19th, and 23rd spokes of the wheel) fall within three main categories:

1. Prime numbers,

2. Squared primes, and

3. Quasi-prime numbers.

Quasi-prime is a new category of numbers defined as the product of semiprimes and/or primes bigger or equal to 5. They include numbers such as 55, 175, 245. Notice that they are all odd numbers and are factorized by numbers residing on the prime-moduli only. Additionally, quasi-primes possess a major distinction from semiprimes in that they exclude the numbers 2 and 3 as one of their prime factors. In this sense, they possess certain properties similar to semiprimes, not being actual primes themselves; still, they retain a degree of *primeness*.

An in-depth analysis of the 24-distribution wheel reveals that the numbers on the prime moduli are *self-contained within these moduli*. In other words, one can form a multiplication grid (*Q-grid*), like the one shown below, where the horizontal and vertical lines consist of prime numbers and semiprimes, e.g., 7, 11, 25, 35, etc. The resultant grid contains all the numbers within the prime moduli that are not prime. Consequently, any number that resides within the prime moduli, and is not contained within the multiplication grid is certainly a prime.

Notice that the diagonal numbers of the grid work as a mirror, with those numbers on top being mirror reflections of those below. Additionally, this line's digital roots follow a repeated sequence of 174471 174471..., made from the numeric group of [1, 4, 7] only. Thus, theoretically speaking, the above Q-grid enables us to test the primality of any number with 100% accuracy and without the need to implement complicated techniques. It also allows us to predict new primes by simply filtering those numbers residing within the prime-moduli while not appearing as products within the Q-grid. It also enables a faster prime factorization, which we explore next.

	5	7	11	13	17	19	23	25	29	31	...
5	25	35	55	65	85	95	115	125	145	155	...
7	35	49	77	91	119	133	161	175	203	217	...
11	55	77	121	143	187	209	253	275	319	341	...
13	65	91	143	169	221	247	299	325	377	403	...
17	85	119	187	221	289	323	391	425	493	527	...
19	95	133	209	247	323	361	437	475	551	589	...
23	115	161	253	299	391	437	529	575	667	713	...
25	125	175	275	325	425	475	575	625	725	775	...
29	145	203	319	377	493	551	667	725	841	899	...
31	155	217	341	403	527	589	713	775	899	961	...
...	...	...	...	...	...	...	...	...	...	...	...

Table 4: The Q-grid for numbers from 5 to 31.

Primality Testing and Prime Factorization

Prime factorization, finding the prime factors of a certain number, is not a complicated problem in principle. However, when it comes to huge numbers, the process becomes very difficult. This is why semi-prime numbers are implemented in many modern encryption techniques. Therefore, factorization is considered a very important problem as by making semiprimes easier to factorize, we render many encryption solutions useless, which jeopardize a huge sector of online businesses and personal information.

There are many methods to test the primality of numbers. The simplest of which is the *trial division* method, where if the tested number n is found to be evenly divisible by any prime number from 2 to $\sqrt{n}$, then it is, by default, not a prime, and vice versa. Other methods are considered more probabilistic than exact, with the tested numbers subjected to some rigorous criteria, and if they pass, then they are considered most probably prime, such as the Fermat primality test combined with Fibonacci or Lucas primality tests.

The Q-grid provides another exact method that can reduce the time for primality testing considerably. To do the testing, we start by checking whether a number n passes the initial primeness-criteria: being odd, having a last-digit of [1, 3, 7, or 9], and having a digital root not equal to [3, 6, 9], etc. Once it passes, n is easily tested as to whether it falls along any of the prime moduli of the chosen s-sides polygon or not. So, for $s = 6$, the prime should fall in either modulus 1 or 5. For $s = 24$, as is the case we are considering, the prime-moduli are: 1, 5, 7, 11, 13, 17, 19, and 23, etc. By taking into consideration all the above, we have reduced the dimensionality of the problem considerably, selecting eight moduli only from a total of twenty-four, eliminating in the process 2/3 of the numbers' space that requires searching. Next, we proceed to the final step, to check whether the number belongs to the Q-prime grid or not. If the original number n is found within the grid, then it is not a prime. Otherwise, it is prime by definition.

Now, if the number n resides along or close to the reflection line of the Q-grid, then primality confirmation is simple. We only need to look for numbers at or close to $\sqrt{n}$ on both axes of the grid. These two numbers will define two small strips that intersect on a small area where we need to look for this number. Beyond this area, numbers either get larger or smaller than n. On the other hand, if the number n is far from the reflection line, then solving the problem requires more complex analysis. Thus, devising algorithms that can search for these numbers without the need to calculate large portions of the Q-grid is an important step in refining and optimizing primality confirmation results.

Similar to the case of primality testing, performing prime factorization using the Q-grid method requires the implementation of a search algorithm that is programmed based on the search criterion we mentioned earlier, along with other supporting conditional statements that exploit the geometry of the Q-grid. The problem's geometrical aspect will be explained further in Part III, where we discuss the many interesting and novel methods we have discovered in the process, which in return, rewarded us with so much insight into not only this specific problem but so many aspects of the physical reality as well.

Quadrant Symmetries

Aside from its primality and factorization capabilities, the 24-based icositetragon wheel and its prime moduli projection, the Q-grid, provide an interesting insight into the relationships numbers have with each other. By projecting the numbers of the eight prime moduli onto the Q-grid, shown below, we discover some interesting behavior. Notice how those Quasi-primes belonging to the 1st, 4th, 5th, and 8th prime moduli continue along the grid without interruption (black lines), while those of the 2nd, 3rd, 6th, and 7th moduli (dashed lines) are interrupted either by each other or by the other four prime moduli. Thus, there is a sense of symmetry breaking between the horizontal and vertical prime moduli that is not obvious from the geometrical symmetry of the icositetragon wheel.

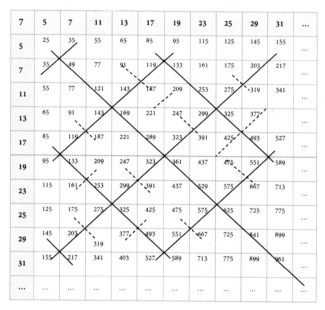

7	5	7	11	13	17	19	23	25	29	31	...
5	25	35	55	65	85	95	115	125	145	155	...
7	35	49	77	91	119	133	161	175	203	217	...
11	55	77	121	143	187	209	253	275	319	341	...
13	65	91	143	169	221	247	299	325	377		...
17	85	119	187	221	289	323	391	425	493	527	...
19	95	133	209	247	323	361	437	475	551	589	...
23	115	161	253	299	391	437	529	575	667	713	...
25	125	175	275	325	425	475	575	625	725	775	...
29	145	203	319	377	493	551	667	725	841	899	...
31	155	217	341	403	527	589	713	775	899	961	...
...	...	...	...	...	...	...	...	...	...	...	...

Table 5: Tracing the prime moduli on the Q-grid reveals symmetry breaking between the North/South (solid lines) and East/West (dotted lines) moduli.

By examining the northern, eastern, and western central axes of the wheel, we find an interesting complementary property. Aided by the figure below, we see that numbers on the east and west moduli add up to every other number on the northern modulus only. For example, numbers 6 and 18 on the horizontal axis add up to 24 on the vertical axes, 30 + 42 = 72, 54 + 66 = 120, and so on. The southern modulus is excluded from this property, which is another asymmetric behavior.

19

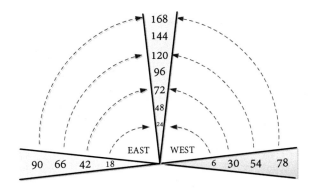

Figure 10: East-west central moduli add up to every other number on the northern central modulus.

Additionally, the northern and southern central moduli complement each other, where two numbers on these moduli add up to produce a number on the southern one only, as illustrated below. So, on the vertical axis we find: 24 + 12 = 36, 72 + 60 = 132, etc. The East-West central moduli, on the other hand, don't exhibit the same behavior.

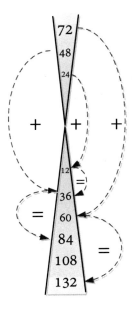

Figure 11: North-south moduli complementary property: adding two numbers from the north-south moduli generates a number belonging to the south modulus only.

Below is an illustration of another complementary property where numbers on two prime moduli add up to a non-prime number on the central moduli. For example, prime numbers 77 and 79 add up to 156, which lies on the southern central modulus. Also, 12 = 5 + 7, 24 = 13 + 11, 24 = 23 + 1, 36 = 19 + 17, 60 = 29 + 31, etc.

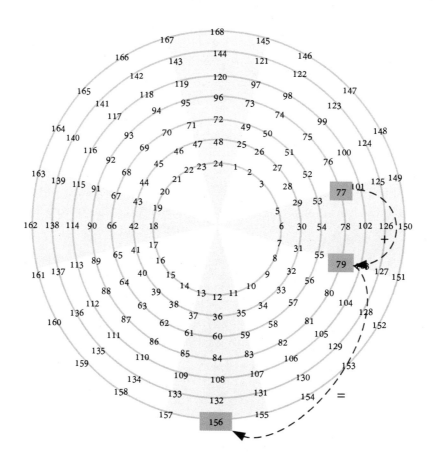

Figure 12: Prime numbers on the prime moduli adding up to composite numbers on the central moduli.

It is obvious from the above analysis that the icositetragon (24-sided polygon) numerical configuration has a quadrant nature. These quadrants are defined by the primary horizontal and vertical central moduli that, at the same time, work as complementary axes where the numbers on the central moduli possess complementary relationships with numbers on other central moduli (Figures 10 & 11), as well as with numbers belonging to the prime moduli (Figure 12) as explained above.

Quadrant symmetry is something we encounter in many fields of physics and mathematics, such as in the complex plane where the real and imaginary axes define a similar configuration, with the quadrants on the negative side often being a continuation or reflection of those on the positive side, as proposed by the complex function continuity theorem. Also, propagating electromagnetic waves define a similar quadrant plane, as we will see later on.

Intrinsic to any coherent quadrant configurations is the existence of a relationship or symmetry between the elements of each quadrant, i.e., having a reflection symmetry around the horizontal axis and/or vertical axes. This is also the case for the numeric icositetragon wheel, as each integer on the central moduli and the prime moduli that surround them possesses *circular-complementary* relationships with numbers on other moduli such that they add up to 360 (or multiples of 360) and are, therefore, parts of the same quadrant integer set. The figure below explains this mirroring relationship between the members of a quadrant set.

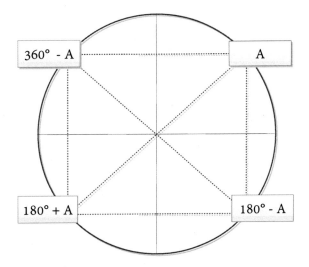

Figure 13: Definition of the mirroring complementary relationship between a set of four numbers with A being a member of the set.

We explain the above quadrant symmetry through an example. Taking number 341 on the 5th modulus, its complementary number 19 is found in the opposite mirrored 19th modulus (5+19 = 24) when reflecting across the central vertical moduli (24 and 12). This is because 341+ 19 = 360. The two additional complements, 161 and 199, are reflected through the horizontal moduli (moduli 6 and 18). Notice how number 199 is complementary to number 19, and so is number 341 to 161. In other words, 199 -19 = 180 = 341-161. These two numbers are part of the same quadrant set not only because 161+199 = 360 but also because they unify the degree and decimal references when they are referenced around a unit circle.

22

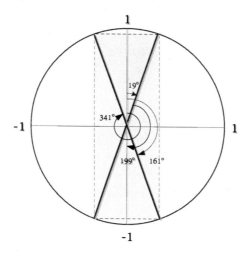

Figure 14: The angular relationships between the four numbers belonging to one complementary set. In this example, 199° is a continuation of 19°, and so is 341° of 161°.

Dividing the decimal numbers by 360 creates a normalized reference for the set where the numbers add up to 1 instead of 360. We can also give these decimal references signs that define their position on the unitary circle and maintain a balance of 0 around the whole circle. This is shown in the figure below for the set [(43, 317), (137, 223)].

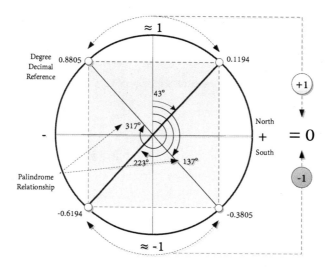

Figure 15: Normalized relationship for one complementary set [(43, 317), (137, 223)] along with the decimal references and signs attributions.

For numbers beyond 360, the same complementary properties apply; however, the numbers will add up to $n \times 360$ where n is an integer equal to 1 for the first 360 numbers, 2 for the next 360 numbers, and so on.

The quadrant relationships are better observed when we look at the trigonometric values of their reference angles, as illustrated in the figure below. The values between the brackets are for the cosine, sine, and tangent of the angles, respectively (*und.* stands for undefined, e.g., irrational values). Notice how they are mirrored across the quadrants, differing only by a $\pm$ sign. There are also complementary symmetries corresponding to central moduli numbers, continuing ad infinitum in multiples of 360 or 180, in a cyclic pattern, as shown below.

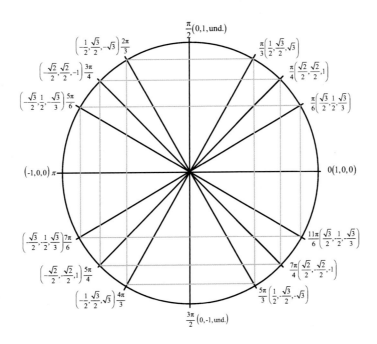

Figure 16: Trigonometric values for three different sets of quadrant-angles mirrored across the four quadrants, differing only with their $\pm$ signs.

All these symmetric and asymmetric numeric relationships are a consequence of the 2-dimensional geometric configuration of numbers and cannot easily be observed in their 1-dimensional linear form. And while the symmetric relationships can be explained via the wheel's geometry, the asymmetric ones cannot be easily explained, especially when they involve the same axes of symmetry.

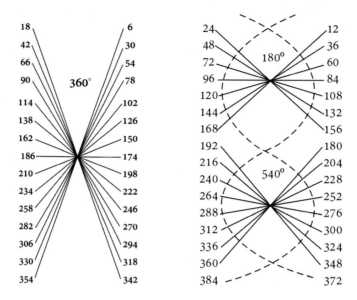

Figure 17: Numbers on the horizontal moduli complete to 360° (left), and numbers on the vertical moduli complete to multiples of 180° (right).

All the above hints to an evident conclusion: that numbers are geometric in nature, and their affinity to circles is most obvious. Their real powers, properties, relationships, etc., are observed and make sense only when they are distributed circularly instead of linearly. This is something we observed early on in the *D*-space, as only when numbers are distributed around circles do their numeric symmetries and relationships become readily apparent and meaningful.

Therefore, it makes no sense to separate numbers and geometry from each other; they should always be studied together as a coherent entity. It is only through this geo-numeric domain that scientific mysteries can be discovered and solved.

'Magical' Numbers

"The wisest and noblest teacher is nature itself."
-Leonardo da Vinci

The Golden Ratio

Throughout the ages, many numbers, whether individual ones, like the golden section, or a set of them, like the Fibonacci sequence, were given a paramount status among all other numbers. They were so revered to the extent that they were labeled magical.

Magic is a word we usually associate with things or events that we cannot explain in a reasonable or scientific manner. It is a process where one thing transforms into another when no such transformation is believed to be allowed. Much of what the ancient believed to be real magic was mostly based on scientific tricks performed by magicians or priests in front of ignorant or uninformed crowds.

Nowadays, magic has become the art of invisible tricks, where the observer is amazed by how the magic trick is done, much more than the magical event itself. In the scientific realm, mathematicians label a number magical if it has certain properties that make it unique but only in a purely mathematical and algebraic sense. On the other hand, a physicist calls a number or set of numbers magical if nature seems to prefer it/them in its design, such as the so-called magical numbers corresponding to the nucleus's energy states.

But even if scientists sometimes call some numbers magical, they do not mean it literally. For them, these numbers are just a consequence of the laws of nature; the *magic* label is just an unintentional misnomer. For the scientists and philosophers of the past, on the other hand, when they observe a pattern, shape, or number that is repeated in nature in different areas and at different levels, they would consider it unique and magical, manifested in nature to reflect the divine design and wisdom. They would try to understand it, imitate it, and use it whenever they could because in doing so, they would be reflecting its divine beauty and purpose within their own work, which would become vibrant and harmonious. In other words, they believed that these numbers would magically transform their earthly creation into a heavenly one.

In simple words, what is widely observed in the natural world and fundamental to its design, whether a number or a shape, is what is considered meaningful and important to the intelligence behind its creation, and consequently, what is considered magical to the ancients. And this is precisely why these numbers are studied here, showing them to be much more magical than ever believed.

There is no other number revered as magical more than the *golden ratio*. Mathematically, it is defined as the cut on a line where the ratio of the larger piece to the smaller one is the same as the ratio of the whole to the larger piece, as shown below.

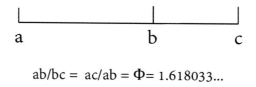

$$ab/bc = ac/ab = \Phi = 1.618033...$$

Figure 18: The golden ratio is found by cutting a line into two segments, where the ratio of the bigger piece to the smaller one equals that of the whole to the bigger piece.

The numerical value of this amazing number (usually referred to as the Greek letter Φ) is the same no matter what the length of the line is, just like number π has the same value no matter what the diameter of the circle is. And similar to most fundamental constants, such as π and e, Φ is an irrational number, a number that continues to infinity without repetition (π and e are even more irrational than Φ, as where Φ can be generated from an algebraic equation, they cannot. These types of numbers are called transcendental.)

The magical status of the Φ constant stems from its unique mathematical properties as well as its ubiquitous presence in the natural world. This is because the golden section Φ, along with its mathematical variations such as $1/\Phi = 0.618...$ and $\Phi^2 = 2.618...$ are embedded within the many natural laws and patterns that govern growth and proportions. For example, Φ is found in the proportions of the human body, such as the face, arms, torso, etc. Every double helix DNA measures 34 angstroms long by 21 wide, with their ratio being 1.619, very close to Φ. Also, the golden section is embedded within the design of every pentagon and pentagram where each of the five small triangles that make up the pentagram (or the large ones that define the pentagon) is a golden triangle, with the ratios of the sides to the base being Φ exactly, as shown below. This fundamental relationship between Φ and the pentagon stems from its mathematical expression, which involves the square root of number 5: $\Phi = (1+\sqrt{5})/2$.

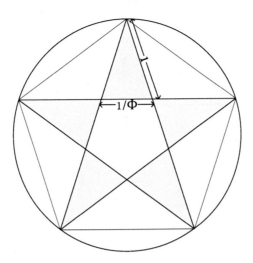

Figure 19: The golden section Φ defines the proportions of every pentagon or pentagram.

If we continuously place golden triangles inside each other, as shown below, and then draw a curve connecting the tips of each triangle, we generate what is known as *the golden spiral*. These types of spirals are called *logarithmic spirals*, defined by maintaining their proportions with every wind. (They are different from Archimedean spirals, which maintain the distances between their arms instead.) The golden spiral is found everywhere in the natural world, from the spiraling forms of many creatures, like the shell of the nautilus, to the horns of some animals, to the spiraling arms of galaxies, etc.

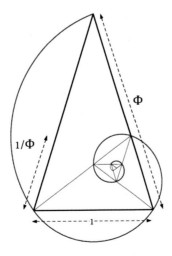

Figure 20: The golden spiral formed from connecting the tips of adjacent golden triangles, all defined by the golden section Φ.

Another geometrical aspect of Φ is the *golden rectangle*, where the ratio between its two sides is Φ, as shown below, left. And similar to the golden triangle, we can trace a golden spiral within embedded golden rectangles.

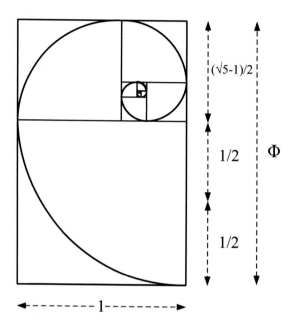

Figure 21: A golden rectangle with dimensions b and c and where c/b = Φ. It nests a golden spiral created from embedded golden rectangles.

Many ancient civilizations, like the Egyptians, the Greeks, and the Mesopotamians, etc., implemented Φ in their architectures and drawings by incorporating the golden rectangles within their works. This is because whenever a painting or a monument is designed with Φ in mind, it will look harmonious and more pleasing to the gazer.

One example is the famous Babylonian *Tablet of Shamash*, dating back to the 9th century BC, where the golden rectangle is embedded within the many elements of its design, as shown below.

Many renaissance painters used Φ in their paintings, such as in Leonardo Da Vinci's *The Last Supper* or in Sandro Botticelli's famous work *The Birth of Venus*. Even in music, some composers arranged their pieces based on Φ, believing it to make their music profound and more enjoyable, which include Dufay, Bach, Beethoven, Mozart, Chopin, to mention a few.

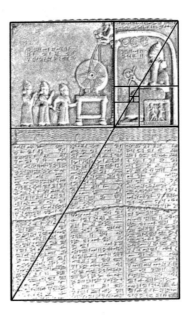

Figure 22: The Babylonian Tablet of Shamash (9th century BC) is designed such that its dimensions reflect the golden rectangle. (Was this an intentional design, or did these dimensions simply resonate with its maker?)

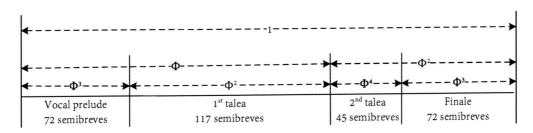

Vocal prelude	1st talea	2nd talea	Finale
72 semibreves	117 semibreves	45 semibreves	72 semibreves

Figure 23: "Vasilissa ergo gaude" of Guillaume Dufay (1397-1474) is almost completely based on the golden section.

So, whether numerically or geometrically, naturally or artificially, the golden section exists in almost everything around us. Its beautiful mathematical and geometrical properties make it the preferred proportion through which nature expresses itself. And later on in the book, when we discuss the wave nature of numbers, we show that the golden section is not only one of the most important constants in nature, but it may very well be the origin of all other constants as well.

30

The Fibonacci Sequence

Fibonacci numbers were known since the time of Ancient Egypt. Their current name is attributed to the man who introduced them to Europe, Leonardo of Pisa, aka Fibonacci (1170-1250 AD). He is reputed to have discovered this sequence of numbers while studying the theoretical growth-rate of rabbits. Starting from a single couple, represented by [1, 1], each successive number is then generated by the summation of the two previous ones, generating a sequence as follows: [1, 1, 2, 3, 5, 8, 13, 21, 34, 55, 89, 144, 233, 377, 610, 987, 1597, 2584, ...]. Note how numbers 21 and 34 that define the scaling of the DNA belong to Fibonacci's sequence.

Interestingly, the golden section and Fibonacci numbers can be generated from each other, as the ratio of each two successive Fibonacci numbers converges to the exact value of the golden section, more so as we move towards larger numbers: 3/2 = 1.5, 5/3 =1.666..., 8/5 = 1.6, 13/8 = 1.625, 21/13 = 1.61538..., 144/89 = 1.61797... Eventually, the ratio converges very closely to 1.6180... (Interestingly, if we take the ratio of the 24th and 23rd Fibonacci numbers, the value matches the golden section string up to 9 digits after the comma: 46368/28657 = 1.618033988... So, we can say that at the 24th number of Fibonacci, we achieve one complete cycle of matching with the golden section.)

In reverse, the Fibonacci sequence can be calculated from the golden section by simple mathematical relations, as shown in the table below. When we remember that Φ is an irrational number while Fibonacci numbers are rational and exact, this correspondence becomes much more interesting.

Fibonacci	Φ
1	$\Phi^0 + 0/\Phi^2$
2	$\Phi^1 + 1/\Phi^2$
3	$\Phi^2 + 1/\Phi^2$
5	$\Phi^3 + 2/\Phi^2$
...	...

Table 6: Fibonacci numbers generated from the powers and reciprocals of the golden section.

The Fibonacci Spiral

The real power of Fibonacci's sequence does not emanate from the sequence's numerical aspect only, but more importantly, from the geometrical pattern its numbers form, the spiral, which is used extensively in nature in arranging and packing its constituents.

A logarithmic Fibonacci spiral can be traced inside a sequence of adjacent squares having Fibonacci numbers for the dimensions of their sides, as shown below. (Notice that this spiral is not the same as the golden spiral as here we have squares in place of golden rectangles.) This geometrical aspect is unique to the Fibonacci sequence. (As we will see in the next chapter, Lucas numbers are generated using the same exact logic used in Fibonacci; however, they do not form geometrical patterns.)

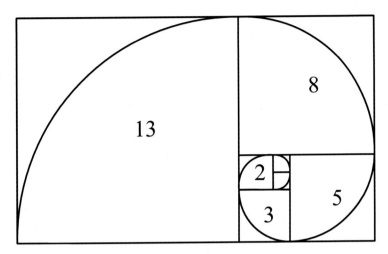

Figure 24: Adjacent squares whose dimensions follow the Fibonacci sequence define a perfect logarithmic spiral.

Plants exhibit Fibonacci numbers in their spiraling forms and in the numbers of these spirals. Artichokes, for example, have clockwise and counterclockwise spirals that are numbered in a Fibonacci sequence: 5 clockwise and 8 counterclockwise. Pineapples also exhibit these numbers in their 3-fold spiraling scales; 8, 13, and 21, and so is the case for the heads of sunflowers, among many others.

When we take the digital root of the Fibonacci sequence, a cyclic pattern emerges with the digits repeating themselves in a 24-fold cycle as follows: [1, 1, 2, 3, 5, 8, 4, 3, 7, 1, 8, 9, 8, 8, 7, 6, 4, 1, 5, 6, 2, 8, 1, 9]. Remember number 24 from the previous chapters, where the 24-moduli wheel exhibited interesting properties, especially in regard to prime numbers. This is another very important place where this number is hidden in plain sight; we only needed the digital root to reveal it.

When these 24 digits are plotted around a circle (shown below), the D-circle exhibits many interesting geo -numeric symmetries. For example, every two numbers on the same diameter complete to 9, along with two groups of 3-6-9 segmentations and their corresponding 3-6-9 D-sums, as shown around the same circle.

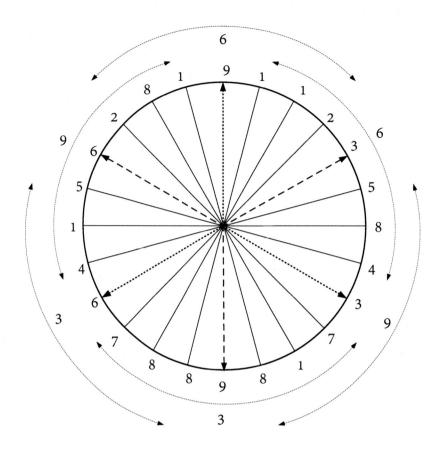

Figure 25: The 24-based *D*-circle of the Fibonacci sequence. Every two numbers on the same diameter add up to 9. Also, the [3, 6, 9] group divides the circle into three segments where the *D*-sum of the numbers inside are also 3, 6, and 9.

The perfectness of these 24 numbers in their geo-numeric symmetries is as elegant as intriguing. If only one number differed from its exact value, the symmetries' wholeness would have been broken. Also notice the perfect triangulation of the group of numbers [1, 4, 7], [2, 5, 8], and [3, 6, 9], with the members of each group, collectively or individually, placed around equilateral triangles. This differentiation of the first nine numbers into these three groups is an observation of utmost importance, which we will repeatedly witness throughout the book.

What more information could Fibonacci's *D*-circle be hiding? And are there other 24-based *D*-circles similar to it? To find out, we look at another famous sequence that shares the same generation methodology with Fibonacci; that of the Lucas sequence.

The Lucas Sequence

The Lucas sequence is another set of numbers that are observed in nature but not as often as Fibonacci's. It has been given its name after the French mathematician François Édouard Anatole Lucas (1842–1891), who is the one that gave Fibonacci numbers their name also. Generating Lucas numbers is very similar to generating Fibonacci's, except that we start with numbers [2, 1] (instead of [1,1]), with each subsequent number being the sum of the two previous ones: [2, 1, 3, 4, 7, 11, 18, 29, 47, 76, 123, 199, 322, 521, ...]. And similar to Fibonacci's, Lucas numbers can be exactly calculated from simple combinations of the golden section Φ and its powers, as shown in the table below.

Lucas	Φ
1	$\Phi^1 - 1/\Phi$
2	$\Phi^1 + 1/\Phi^2$
3	$\Phi^2 + 1/\Phi^2$
4	$\Phi^3 - 1/\Phi^3$
7	$\Phi^4 + 1/\Phi^4$
...	...

Table 7: Generating Lucas numbers from the powers of the golden section.

Lucas numbers are related to many mathematical problems and can be found in the phyllotaxis of some plants but are rare compared to Fibonacci numbers. So, for example, instead of counting 5 clockwise and 8 counterclockwise spirals, we would count 4 and 7, as is the case in the leaf distribution of the Greenovia Aurea, for example.

In the D-space, Lucas numbers exhibit another 24-fold repetition cycle with similar geo-numeric symmetries to those of Fibonacci's: [2, 1, 3, 4, 7, 2, 9, 2, 2, 4, 6, 1, 7, 8, 6, 5, 2, 7, 9, 7, 7, 5, 3, 8].

Interestingly enough, if we superimpose the two D-circles of Fibonacci and Lucas such that numbers 3, 6, and 9 are matched on both circles, and then took the D-sum of the superimposed numbers, we would generate a reverted (rotated by 180°) Fibonacci circle. (A rotation by 180° is synonymous with plotting the numbers in a counterclockwise fashion instead of clockwise.) More interestingly, if we superimposed the two circles after a 180° rotation of one of them in respect to the other (such that the 3, 6 numbers in one of them are reversed compared to the other) and then took the D-sum, the resulting circle would consist of the numbers [3, 6, 9] only, as shown below in figure (27).

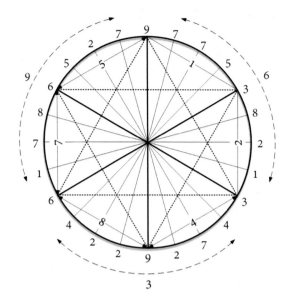

Figure 26: The 24-base circle of the Lucas sequence. Similar to Fibonacci's, every two numbers on the same diameter add up to 9. Also, the [3, 6, 9] group divides the circle into three segments where the *D*-sum of the numbers inside are 3, 6, and 9.

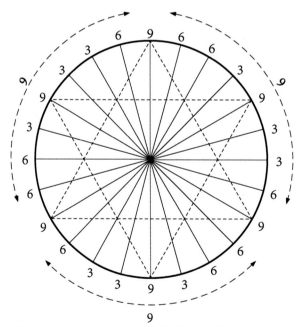

Figure 27: Superimposing Fibonacci's *D*-circle on top of a 180° rotated Lucas one generates an all [3, 6, 9] circle.

The above *D*-circle is interesting as it consists of the sequence [3, 3, 6, 9, 6, 6, 3, 9] repeated three times, as shown below. Moreover, this sequence follows the same by-sum logic used in the other two sequences of Fibonacci and Lucas, but with the initial digits being [3, 3]: 3+3 = 6, 3+6 = 9, 6+9 = 6, and so on.

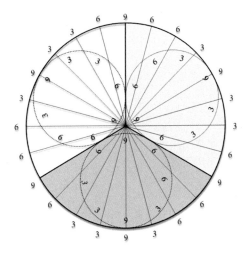

Figure 28: The triple identical 8-based sub-circles of the 3-6-9 *D*-circle. Each sub-circle can be thought of as one full octave.

Each one of the 8-based smaller circles has the interesting property of generating the same ratios between its numbers, but in a different order, whether we move clockwise or counterclockwise. Moreover, all these ratios happened to add up to 10, exactly: 3+1+(1/2)+(2/3)+(3/2)+1+2+(1/3) = 10, which is the *Decad;* Pythagoras's holiest number of the Tetractys.

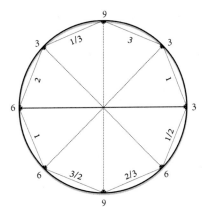

Figure 29: Looking at one sub-circle of the 3-6-9 *D*-circle, the mutual ratios of the numbers all add up to 10. They can be used to create the ratios of the musical octave.

As we saw earlier when we talked about the doubling process of numbers, the [3, 6, 9] group is unique when compared to the other two groups of [1, 4, 7] and [2, 5, 8]. And a circle made up completely of these numbers hints at an interesting dynamic happening between the two sequences of Fibonacci and Lucas. We further investigate this in the next section.

The Magical Quartet

As we have shown above, the Fibonacci and Lucas *D*-circles are generated using the same by-sum logic, differing only by their initial two digits. But how many other *D*-circles can we generate using this same logic? It turned out that whatever two initial numbers we use, we will always get sequences having 24-fold repeated cycles in the digital root space. For example, starting with numbers 3 and 1 (let us call them the seeds), we also generate 24 repeated numbers, which, when plotted on a circle, exhibit a diagonal 9-symmetry and double [3, 6, 9] segmentation as shown below. This sequence of numbers is referred to as Pibonacci, and so we will call its circle the Pibonacci *D*-circle.

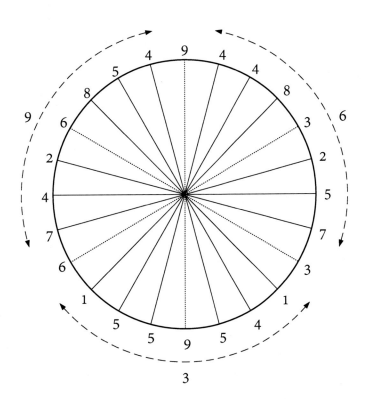

Figure 30: The Pibonacci *D*-circle, generated from a seed of [3, 1], using the same logic of Fibonacci's and Lucas's sequences.

Interestingly, starting with any two adjacent numbers on the *D*-circles of either Fibonacci, Lucas, or Pibonacci, and then using them as the seed of a new sequence, the new sequence will generate the same exact *D*-circle from which the seeds were taken from, taking into account that we are summing in the right direction. For example, starting with numbers [7, 2] on Lucas's *D*-circle, moving clockwise, the next number will be 7+2 = 9, then $D(9+2) = 2$, then 2, 4, ..., until the same original Lucas *D*-circle is generated back again. In other words, these circles have the seeds of their own generation.

But how many of these 24-based *D*-circles are there?

It is not that hard to find out. What we need to do is to write down a table with all the possible combinations for the first two numbers (the seeds), as shown below. Next, we exclude all the pairs that are found in the *D*-circles of Fibonacci, Lucas, and Pibonacci, as these pairs will generate the same *D*-circles back again. These numbers are indicated by the white, dark gray, and gray cells, respectively. Doing so, we find that all the remaining pairs (darker cells) belong to one sequence only, that of the 3-6-9 *D*-circle.

1, 1	1, 2	1, 3	1, 4	1, 5	1, 6	1, 7	1, 8	1, 9
2, 1	2, 2	2, 3	2, 4	2, 5	2, 6	2, 7	2, 8	2, 9
3, 1	3, 2	3, 3	3, 4	3, 5	3, 6	3, 7	3, 8	3, 9
4, 1	4, 2	4, 3	4, 4	4, 5	4, 6	4, 7	4, 8	4, 9
5, 1	5, 2	5, 3	5, 4	5, 5	5, 6	5, 7	5, 8	5, 9
6, 1	6, 2	6, 3	6, 4	6, 5	6, 6	6, 7	6, 8	6, 9
7, 1	7, 2	7, 3	7, 4	7, 5	7, 6	7, 7	7, 8	7, 9
8, 1	8, 2	8, 3	8, 4	8, 5	8, 6	8, 7	8, 8	8, 9
9, 1	9, 2	9, 3	9, 4	9, 5	9, 6	9, 7	9, 8	9, 9

Table 8: The table of all possible seeds to be used in the by-sum Fibonacci sequence logic.

So, in total, we have four circles that are generated using the same mechanism, all of which repeat in a 24-fold cycle. I will call these four circles the *Magical Quartet*.

What is interesting about these circles is that all four of them can be generated from each other by superimposing one on top of the other and adding the overlapped numbers, as shown in the table below.

+	Fibonacci	Lucas	Pibonacci	3-6-9
Fibonacci	Inv.[Lucas]	Inv.[Fibonacci]	Inv.[Pibonacci]	[Lucas]
Lucas	Inv.[Fibonacci]	Inv.[Pibonacci]	Inv.[Lucas]	Pibonacci
Pibonacci	Inv.[Pibonacci]	Inv.[Lucas]	Inv.[Fibonacci]	Fibonacci
3-6-9	Lucas	Pibonacci	Fibonacci	Inv.[3-6-9]

Table 9: The different combinations of the magical quartet and the additions outcomes. (Inv. stands for inverted circle; 180° rotation.)

As it is obvious from the same table, all these *D*-circles are connected to each other via the 3-6-9 one, which has the ability to transform one circle into the other, as shown below.

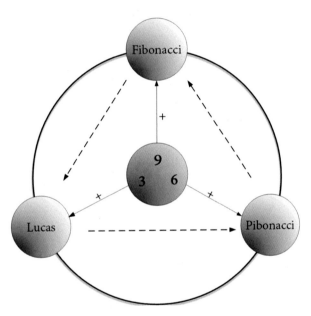

Figure 31: The transformation of the magical quartet. Adding one magical *D*-circle to another generates a third. The [3, 6, 9] *D*-circle has the ability to transform the other three *D*-circles into each other.

Moreover, adding the three *D*-circles together (Fibonacci + Lucas + Pibonacci) generates an inverted 3-6-9 circle, and when all four of them are added together, an all-9 *D*-circle is generated, as shown below. Therefore, they all complete each other.

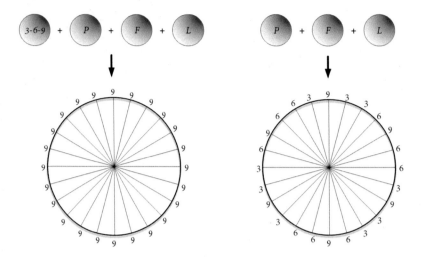

Figure 32: Adding the magical quartet members together generates an all-nines *D*-circle. While adding them together, without the [3, 6, 9] *D*-circle, generates an inverted [3, 6, 9] *D*-circle.

We can create a numeric hierarchy from the three *D*-circles of Fibonacci, Lucas, and Pibonacci, as shown below. Notice how the numbers in the radial direction are ordered such that each branch is made entirely from the members of one group of the triplet [1, 4, 7], [2, 5, 8], and [3, 6, 9].

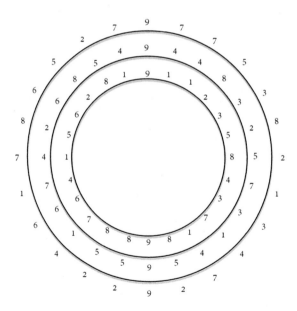

Figure 33: Nesting the three *D*-circles of Fibonacci, Lucas, and Pibonacci inside each other generates a hierarchy where diagonal numbers follow the triplet of [1, 4, 7], [2, 5, 8], and [3, 6, 9], either collectively or individually, e.g. [3, 3, 3].

Thus, the magical quartet *D*-circles are not only perfect individually but also collectively. Together, they create a complete space of 24-ness that cannot be achieved outside the digital root. The 3-6-9 *D*-circle works as an intermediator between the other three, able to transmute one circle into the other, like a philosopher stone or a *philosopher circle* for that matter. Understanding the numeric composition of these *D*-circles and their entanglements will definitely improve our understanding of the numbers behind them. These numbers are unique not only for their digital root sequencing or their natural occurrences but also for their generation logic.

Most of the number sequences are purely mathematical in the sense that they can be represented by mathematical formulas that depend on a certain variable, *x*; by substituting any number for *x*, a new number is automatically generated. For example, pentagonal numbers can be calculated from the mathematical formula of $P_x = 0.5 \times (3 \times x - 1)$ where $x = 0, 1, 2, \dots \infty$. However, Fibonacci, Lucas, Pibonacci, and even 3-6-9 sequences are not that perfectly mathematical. Behind their simple logic lies a fundamental difference that sets them apart from those perfectly systematical numbers. To illustrate why it is the case, we need to scrutinize their generation logic, step by step.

Mathematically, these sequences can be generated using the simple expression: $U_{n+1} = U_n + U_{n-1}$, where each new number is generated by summing the previous two. The difference between these numbers and the other ones, e.g., pentagonal numbers, lies exactly in this specific step; the need to know the present number (U_n) and the previous one (U_{n-1}) to produce the next (U_{n+1}). Hence their generating mechanism depends on the past and present in order to produce the future. Pentagonal numbers, for example, do not need to know their present nor history in order for them to exist.

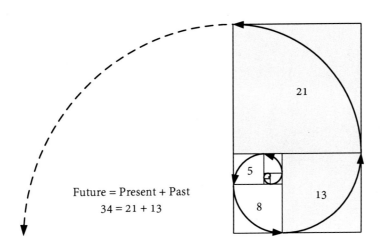

Future = Present + Past
34 = 21 + 13

Figure 34: Generating any number of the magical quartet requires knowing the history and present to create the future.

However, knowing the history requires a feedback loop and a memory, a consciousness. It is our consciousness that allows us to look at the previous two numbers to be able to generate the third. They are consciousness-requiring numbers.

Beyond these conscious numbers lies the realm of chaos, where numbers exist without any perceived pattern or known mathematical formalism. And so, between these two extremes of numbers: the completely systematic and mathematical, and the completely random and chaotic, lies the consciousness numbers of the magical quartet, the most D-space symmetrical ones, the numbers of life.

"The main purpose of science is simplicity, and as we understand more things,

everything is becoming simpler."

-Edward Teller

The Wave Dynamic of Numbers

"It is seen that both matter and radiation possess a remarkable duality of character, as they sometimes exhibit the properties of waves, at other times those of particles. Now, it is obvious that a thing cannot be a form of wave motion and composed of particles at the same time – the two concepts are too different."

-Werner Heisenberg

All Is Wave: Even Numbers

Waves are one of the most studied natural phenomena, being relevant to most scientific fields. This is due to their ubiquitous presence everywhere in the universe, from the macrocosmic to the microcosmic levels.

The most common forms of waves are those observed in water, mostly due to the mechanical forces that winds exert on water surfaces like seas, lakes, ponds, etc. Or, simply drop a stone into still water, and circular waves will be generated, expanding outwardly from the center (where the stone was dropped). Sound is also described as propagating wave-forms that travel by compression and expansion of air molecules.

Waves emanating from two or more sources interfere with each other creating a unique pattern of alternating regions resulting from the constructive and destructive interference between the troughs (highest points) and crests (lowest points) of the waves. When troughs meet troughs (or crests meet crests), they add up constructively, while when a trough and crest meet, they add up destructively, canceling each other out.

Interference patterns are the hallmark of wave phenomenon and one of the essential tools used by scientists to measure the interfered waves' properties, such as their frequencies, their speeds, and their distances, among many other properties. In fact, the interference pattern is one main proof of the phenomenon's wave-like nature, such as light, as demonstrated by Young's famous double-slit experiment, where the interference of two point-like sources of light creates alternating regions of light (constructive interference) and dark (destructive interference).

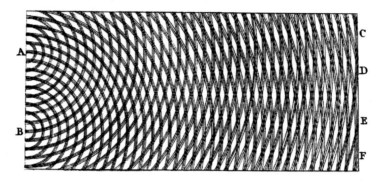

Figure 35: A sketch by Thomas Young illustrating light diffraction-interference pattern.

At the beginning of the 20th century, it became evident that the centuries-old classical models used to explain physical phenomena were inadequate in explaining the new observations and discoveries the experimental physicists were making. A breakthrough came in 1913 when Niels Bohr, a Danish physicist, suggested a novel way to look at the orbits of the electron around the proton in the hydrogen atom; instead of a continuum of possible energy states, only discrete or quantized ones are allowed. The success of his model in predicting the atomic spectrum of the hydrogen atom ushered in the age of quantum mechanics.

Another breakthrough came from the French physicist Louis De Broglie who suggested that particles may behave like waves inside the atoms, exhibiting wavelengths and frequencies, just like sound and light do. This strange hypothesis explained many observations and opened the door into an exciting new reality; however, not an easy one to comprehend nor to visualize. The wave-particle duality was confirmed in 1927 when a different type of Young's double-slit experiment was performed, the Davisson-Germer experiment, where interference diffraction patterns were observed from a stream of electron particles instead of light. And previously, in 1905, Albert Einstein already suggested in his photo-electric paper that light is made of discrete wave packets of energy called *photons*. Thus, light also behaves either as a wave or as a point-like entity, depending on the experiment performed and its observation method. Consequently, the term wave-*particle duality* was coined to describe this mysterious phenomenon.

But where does this duality come from? And how deep does its ramification reach? We definitely can't feel it at large macrocosmic levels. We need to probe to the tiniest levels to detect its origin, tinier than the fundamental particles themselves. There are many levels below that of the elementary particles. For example, protons and neutrons are believed to be made of even smaller entities called *quarks*, three of them in each. And in String Theory, everything, including quarks, is supposed to be made of tiny vibrating strings, which is the most fundamental level physicist reached in their theories.

But there is still one level, more fundamental than the tiniest strings: that of numbers themselves, the numbers that construct the physical and mathematical constants which govern space and natural forces, such as π, e, the fine structure constant, etc. It is the simplest and most abstract level one can think of, the initial frontier.

The Particle Aspect of Numbers

The earliest recorded mention of the aspect of atoms goes way back to the 5th century BC, where the Greek philosopher Leucippus believed matter to be continuously devisable until one reaches a point where it cannot be divided anymore, which is the size of atoms. By the beginning of the 20th century, atoms were found to be made of even smaller entities, called *elementary particles*, mainly the electron, the proton, and the neutron. Consequently, the field of quantum mechanics emerged, forcing physicists and mathematicians to develop novel mathematical frameworks and tools to enable them to describe these particles and their interactions.

In order to describe the quantum states of subatomic particles and to model their interactions, physicists assigned them specific numbers called *Quantum Numbers*. These numbers are conserved and dictate how the particles behave and whether a certain reaction is allowed or not. For example, in beta decay, where a neutron decays into a proton, an electron, and an antineutrino, the quantum charge number should be conserved before and after the interaction: $n^0 \rightarrow p^+ + e^- + \bar{v}$. As the proton is considered positive, the electron negative and the antineutrino carries no charge (just like the neutron), the equation transforms into the following digital identity: $0 = 1 - 1 + 0$. There are many other qualities assigned to particles, such as angular momentum, energy states, spin, etc. Most of them are described by sets of integers, either discrete, e.g., [-1, 0, 1] (as in the charge number, we just saw) or continuous, as in principle quantum numbers [1, 2, 3, ...].

But we are not constrained to use these specific numbers only. We can use different sets of numbers, such as from 1 to 9, to describe the elementary particles and their behaviors. However, this requires the implementation of digital root math. For example, if we assign to particles the numbers shown in the table below, we find that these numbers also satisfy the beta decay and any other elementary reaction.

The original idea behind this table came from the Vedic Square table's shapes, explained earlier in chapter 2, where we showed how those shapes corresponding to numbers adding to 9 were mirror images of each other. Number 9 will occupy the position of a photon or, in broader terms, energy, as photons are believed to be their own mirror images, just like the shape formed by number 9 on the Vedic Square corresponded to a square (or a dot), which is its own mirror image. Similarly, each particle is believed to have its own mirror image, like the electron and positron, proton, and antiproton, etc. Moreover, when a particle and its antiparticle are combined, they create a burst of pure energy (number 9), just like 1+ 8, 2+7, 3+6, and 4+5 do.

Particle	Anti
$n^0 \rightarrow 1$	$n^{0-} \rightarrow 8$
$p^+ \rightarrow 2$	$p^- \rightarrow 7$
$e^- \rightarrow 3$	$e^+ \rightarrow 6$
$v \rightarrow 4$	$v^- \rightarrow 5$
Energy (or photon) is 9	

Table 10: Assigning numbers from 1 to 8 for the subatomic particles and antiparticles. Number 9 will take the place of photons or energy.

We can check the above table by looking at some fundamental reactions of the subatomic world: the β^- decay, the β^+ decay, and the electron capture (K-capture), and see whether the above numbers satisfy these reactions or not.

As explained above, the β^- decay is the spontaneous decay of the neutron into the three subatomic particles: the proton, the electron, and the antineutrino, which is responsible for the radioactive behavior of some elements. It goes as follows:

$$n^0 \rightarrow p^+ + e^- + v^-$$

When we substitute the numbers of the above table into this equation, keeping the digital root in mind, we get the following consistent result: $D(1) \rightarrow D(2 + 3 + 5) = D(10) = 1$. Now we try the β^+ decay, which is the energy-induced decay of the proton and goes as follows:

$$energy + p^+ \rightarrow n^0 + e^+ + v$$

D-spacing gives: $D(9 + 2) = 2 \rightarrow D(1 + 6 + 4) = D(11) = 2$. Again, working perfectly. In the (K-capture) process, a proton will absorb an electron and turn into a neutron, as follows:

$$energy + p^+ + e^- \rightarrow n^0 + v$$

And again: $D(9 + 2 + 3) = 5 \rightarrow D(1 + 4) = 5$.

So, all the numbers worked out perfectly in this D-space version of the atomic world.

This particular D-space number assignment is chosen such that each particle, when added to its antiparticle, results in number 9, which is the zero of digital root math, as we showed earlier. This coincides with the usual assignment of quantum numbers, where adding 1 and -1 results in 0. However, number 9 is not 0 or nothingness, and this is where the two systems differ, as when a particle and antiparticle unite, they annihilate each other in a burst of pure energy, and this is what number 9 is telling us here: that the unification results in energy, not 0 (like numbers 1 and -1 would have produced). Moreover, when a proton and electron unite, whether as a dipole or an atom, they do not reduce to zero, as 1-1 will produce. In our numeric regime, a proton and an electron will produce the number 5, indicating unity and marriage, as in male (3) and female (2). Thus, the *D*-space number assignment is more accurate , more complete, and more informative than the mainstream one. Additionally, nothingness doesn't exist in the *D*-space, only nines. And number 9 stand for energy. Therefore, the vacuum is not empty but instead filled with energy, matching what modern physics is suggesting about the vacuum being filled with what is called *zero-point energy*.

The above perfect particles/numbers correspondence indicates that numbers can enact the role of particles perfectly (and energy as well). They even dictate how particles interact and behave. In a sense, it is as if from the properties of numbers that the physical reality inherits its fundamental laws. However, as we explained earlier, the point-like aspect is not the only one that particles exhibit; they have a wave-like nature that is as fundamental as the former. This wavy aspect is described by specific equations, such as Schrödinger's equation, and interpreted as a probability density spread over spacetime, communicating a physical meaning only when its complex value is squared. Are numbers up to the task? Do they exhibit a wavy nature as well?

The Wave Aspect of Numbers

Natural numbers [0, 1, 2, ..., ∞] only increase in magnitude, which don't exhibit any wave-like behavior. However, when numbers from 1 to infinity are distributed, let's say within three columns, we find that their digital roots expose a wavy pattern embedded underneath, as shown in the table below.

NATURAL NUMBERS			DIGITAL ROOT		
1	2	3	1	2	3
4	5	6	4	5	6
7	8	9	7	8	9
10	11	12	1	2	3
13	14	15	4	5	6
...	...	...	...	...	...

Table 11: If divided into three columns, the digital roots of natural numbers repeat in the triplet group of [1, 4, 7], [2, 5, 8], and [3, 6, 9]

The 1st column of the digital root repeats in the sequence [1, 4, 7], which resembles a wave, as shown below, and so is for the other two columns, [2, 5, 8] and [3, 6, 9].

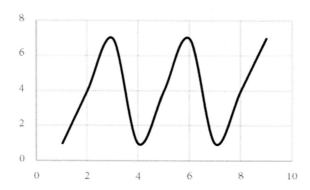

Figure 36: A wave-like pattern made from the repeated pattern of [1, 4, 7]. The wavelength of this pattern is four digits long.

Numbers can be either integers, as we saw above, with no commas or leftovers, or non-integers (floats), like 5.65, or 2.7182818... For the case of floats, there are several types. There are those that are rational, having few numbers for their remainders or where the remainders repeat indefinitely in a specific pattern, like $1/7 = 0.142857\ 142857\ ...$, and there are irrational numbers, being the solution of some algebraic equation, such as the golden section Φ, whose remainder continues to infinity in a completely random fashion. On the far extreme, we have transcendental numbers, such as π and Euler number e, with their remainders also infinitely random; however, unlike irrational numbers, they cannot be expressed by any mathematical expression nor solve any known algebraic equation. These infinitely long numbers, such as $1/7$, π, e, etc. can also be thought of as digital waves, propagating through the infinite numeric space.

The rational value of $1/7$ is of particular interest as it perfectly captures the essence of a wave. The infinitely long pattern of its remainder mimics a wave when plotted individually and more so collectively, as shown below.

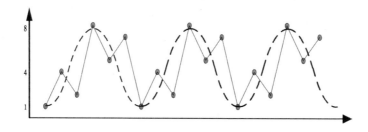

Figure 37: The wave-like pattern of the remainder of 1/7. The plot illustrates three repeated cycles of [142857].

In fact, taking the inverse or reciprocal of numbers is the simplest mathematical method to create float numbers with repeated patterns, e.g., 1/6 = 0.16666..., or 1/137 = 0.0072992700 72992700..., etc. Most of these repeated digits form wave forms similar to the one we saw for 1/7, if not better, as shown below for 1/273 = 0.003663 ...

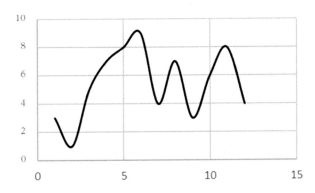

Figure 38: The remainder of 1/273 exhibits a perfect wave-like pattern. The plot illustrates two repeated cycles.

We will thoroughly investigate the reciprocal operation later on. For now, we can safely conclude that numbers do observe wave-like patterns. We can even say that numbers carry certain wavelengths (or frequencies) either individually, as in their repeating remainder, or collectively, as in their digital roots. However, if this numeric wave-like essence is by any means related to the one we observe everywhere in nature, then it should exhibit some of its physical properties as well. And there is no property of waves more fundamental than interference.

Numbers' Interference

As mentioned earlier, interference patterns emerge when two or more waves of the same nature interfere with each other. The resulting waveforms either increase or decrease in amplitude depending on many factors, such as the frequencies of the two source waves, their wavelengths, their separation, speeds, phase, etc.

When the two waves are in phase, their amplitudes add up, and they interfere constructively. When they are out of phase, their amplitudes subtract, interfering destructively, as shown below.

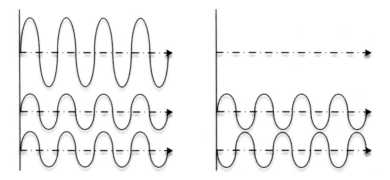

Figure 39: The interference of two waves having the same amplitude. Left: When in phase, the two waves interfere constructively, resulting in a wave of larger amplitude. Right: When they are 180° out of phase, they interfere destructively, canceling each other altogether.

Of the most basic interference patterns is the one resulting from two point-like sources, each producing propagating circular waves, as shown below. The source can only generate propagating waves within a material or medium in which it causes a disruption of some sort, such as water. The source could be of similar nature to the wave medium or of a different one. For example, we can generate waves on the surface of the water using stone pebbles or using droplets of the same water. So basically, the source and the medium of the wave phenomenon can be of the same origin, which is an important concept to bear in mind.

We already know how to generate numeric waves from the inverse or reciprocal operation, such as in 1/7. Thus, we can think of the $1/x$ operation as if number 1 is producing a disruption into the medium defined by the number x. And just like some materials produce better waves than others, some numbers produce better numeric waves than other numbers. So, for $x = 7$, the medium is considered very good for wave propagation, with $1/7 = 0.142857\ 142857\ ...$, while for $x = 2$, the reciprocal operation produces 0.5, which is not a wave. (We can think of it as a local disturbance that doesn't propagate, a sort of permanent dint in the medium, for example.)

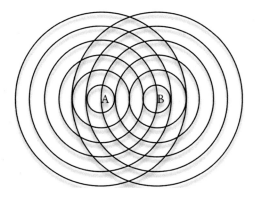

Figure 40: The interference of circular waves produced by two point-like sources, A and B.

And while the repeated pattern of 1/7 is six digits long, for other numbers, the pattern repeats in much larger cycles. For example, for prime numbers, the cycle is most often on the order of the prime -1 (e.g., for p = 19, the remainder repeats in a cycle of 19 - 1 = 18 digits long). In this sense, prime numbers are one perfect medium for generating numeric waves. (Remember that 7 is also a prime whose cycle is 7 - 1 = 6.) There are some anomalies, of course, like the famous prime constant 137, whose wavelength is eight digits long, 1/137 = 0.0 07299270 07299270… However, it produces an even better-looking waveform than 1/7, as shown below.

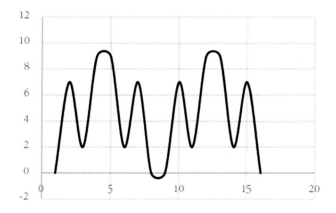

Figure 41: The wave pattern produced by the reciprocal of prime 137, whose remainder repeats every 8 digits. The figure illustrates two cycles.

Looking at the above figure, it is obvious that this wave is not sinusoidal. These types of *modulated* waves can be thought of as carrying information, which is somehow expected from this very important constant of nature (but what could this information be?).

Now, what will happen if two of these numeric waves meet? Will they also interact in a constructive or destructive manner?

One interesting property of repeating patterns emanating from the reciprocal phenomenon is that most of them will complete each other to 9. Take 1/7 as an example; the 142 part, when added to 857, results in 999. Remembering what we argued earlier, using the *D*-space logic, that number 9 works just like a 0, this particualr result of triple 9 makes a lot of sense when compared to regular waves, where adding the crests and troughs of the same sinusoidal wave results in the annihilation of both.

Another example is 137, whose remainder repeats in the sequence of 07299270; adding 0729 to 9270 results in 9999 as well. (This is, in a sense, similar to adding a particle to its antiparticle. Hence, the nines here may suggest energy, not nothingness, as it is already known in physics that energy is never lost; it only transforms from one form to the other. And waves are nothing but energy, materialized in a wave-form.)

However, not all numbers' repeated reciprocals add up to 9. Consider number 78, a non-prime number whose reciprocal repeats in [128205]; adding 128 to 205 results in 333. Even though $D(333) = 9$, still, this is not the same as the expected 999. The graph of this number reciprocal explains why. As shown below, this wave is neither sinusoidal nor symmetric around its half-cycle, and therefore, the crest and trough do not cancel each other out.

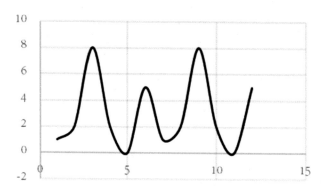

Figure 42: The wave pattern produced by the reciprocal of 78, and for three cycles. Notice the wave pattern is not sinusoidal nor symmetric. These types of waves do not add up to nines.

The above anomaly is not the case for all non-prime numbers. Take 1/76, for example, whose period repeats in an 18-digits wavelength of the numbers [315789473684210526]. Both halves of the wave add up to nines: 315789473 + 684210526 = 999999999. Even though its graph is not that sinusoidal, still, one half is the mirror image of the other (reflected around the horizontal plane). Therefore, they are able to cancel each other (complete to 9).

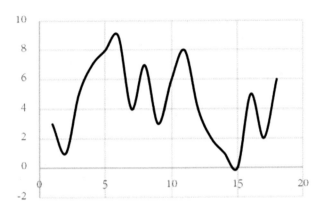

Figure 43: The wave pattern, produced by the reciprocal of 76 and for one cycle. Notice how one half of the wave is but a flip of the other half.

So why do some numbers have reciprocals that complete to nines and others don't? Clearly, this has to do with the shape of the wave, as when half the wave is a reflection of the other one, we have a completion, while when they are not, we don't. Maybe the question should be rephrased as to why some numbers have mirrored wave halves while others do not?

One criterion for those anomalies is that the digital root of the denominator $D(x)$ (the medium) is either [3, 6, or 9]. For example, as we saw above, number 78 was one such anomaly, and its digital root is 6. Another anomaly is number 42, whose reciprocal repeats the sequence 238095, and 238 + 095 = 333. Also, for number 129, whose digital root is 3, the repeating pattern is 775193798449612403100 having an odd wavelength of 21 digits. Not only are we unable to generate nines from this sequence, but also we can't split it into two halves. But not all numbers whose digital roots equal to 3 have odd wavelengths. Numbers like 147 and 138, for example, have even periods that do not complete to 9. Also, number 945 repeats in the sequence of 105820, whose digital root is 7, not even close to 9: 105 + 820 = 925 and $D(925) = 7$. On the other hand, all quasi-prime numbers >5 have reciprocal values that sum to 9 (or digital root of 9, of course).

Thus, one way to understand the problem of having asymmetric waves is to look at the digital root of the number generating the wave. If the digital root is equal to [3, 6, 9], then the waveform is asymmetric. This condition, as we remember, is satisfied only by numbers that are explicitly non-prime. For other digital roots, if the number is not prime, there is a big chance for the wave to be asymmetric.

As we have observed, over and over again, numbers and geometry, especially circular geometry, coalesce in perfect harmony. With their union, the very best of the two is brought up in a compact image that is meaningful as well as very comprehensible. Similarly, the completion of the remainder (or not) is best illustrated around a circle. As shown below for the case of 1/7, the numbers that add up to 9 occupy the same diameter.

We can think of each pair of numbers on the same diameter being 180° out of phase of each other. For those non-completing reciprocals, the pair of numbers on both sides of the same diameter will not add up to 9, while in the case of an odd period, no two numbers occupy the same diameter to start with.

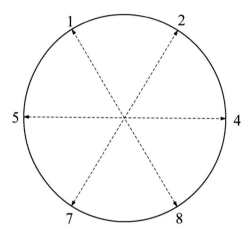

Figure 44: The septenary numbers (1/7), when plotted on a circle, align such that every two numbers occupying the same diameter add up to 9.

In summary, numbers do form numeric waves that interfere as regular waves do. This indicates that numbers express the same particle-wave duality we found in the quantum field of light and matter. And just like numbers control the particle aspect of matter, they may control its wave aspect as well. This we investigate next, where we delve deeper into this phenomenon, trying to fathom its consequences and ramifications.

The Interference Matrix of Numerical Waves

In this chapter, we take a closer look at the interference pattern of waves. We start with two point-like sources creating circular wavefronts, as shown below. Please note that we are considering theoretical mathematically-perfect sources that would create perfect circular waves that do not change their forms with propagation.

Zooming into the pattern, we find the circular waveforms converging on specific points, which we call nodes and numbered 0, 1, 2, etc. According to the Huygens-Fresnel principle, each one of these nodes can be considered a separate point-like source, producing propagating circular waves.

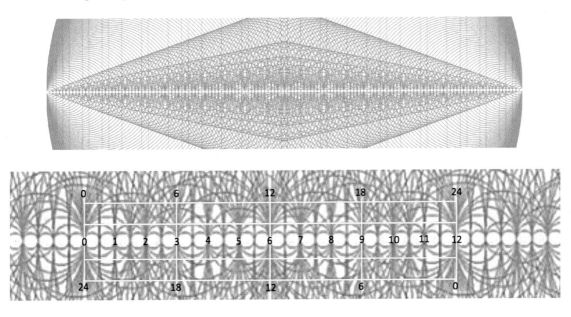

Figure 45: Top: Interference pattern from two point-like sources. Bottom: The same interference pattern, magnified to illustrate the fractal nature of the pattern. We can see that the morphology of the wave patterns (both on top and on the bottom below a horizontal plane) appear to repeat every 24 cycles. This is consistent with the Fibonacci series analysis in *D*-Space

By comparing the pattern of the converging lines at each source point, we sense those lines emanating from the 0th node are identical to those emanating from the 6th and the 12th ones, as if the pattern is obeying a periodicity of 6 units, depending on the level of the fractal. This could be because the pattern is built on a doubling principle where each circle encompasses a number of smaller circles inside it in a hierarchal doubling fashion, starting from two circles (2, 4, 6, 8, ...) or three circles (3, 6, 12, …). And because the number 6 is divisible by 2 and 3 simultaneously, it stands out as the basic unit for this fractal of circles.

Recall from chapter 2, number 6 played the central number in the prime generation formula of $6k \pm 1$. Finding it here, occupying the central position of the wave numeric interference pattern, emphasizes the key role this number plays in both the particle and wave aspects of numbers.

We can also think of the wave pattern, reproduced below, as a numbering or even measuring system, where the nodes set a scale, like a measuring ruler. With fractals built upon fractals of nodes and circles, the ruler's unit can be enlarged or reduced like the units of the metric system: millimeter, centimeter, meter, etc. This wave ruler is one concept that will prove to be of utmost importance later on.

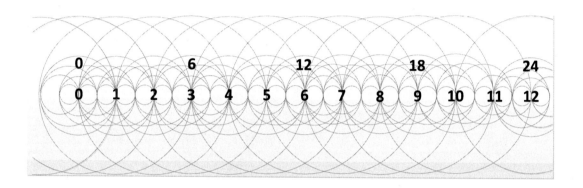

Figure 46: Numeric representation for the nodes of the interference pattern, for a scale from 0 to 12. The pattern can work as a measuring ruler with its unit based on the fractal level of the interference nodes

We further investigate the above pattern by concentrating on the range from 1 to 121, and for reasons explained shortly. As shown in the figure below, the numbers on the central horizontal axis, from 1 to 121, are referenced by their digital roots instead of their original full values for better illustration of the patterns of the wave matrix. On this scale, the central number of the grid is number 60.

Looking at the prime numbers of this range, we notice they exhibit a kind of symmetric distribution around the center number of 60 (for $k = 10$ in the equation $6k\pm1$). First, the two numbers bounding it, 59 (60-1) and 61(60+1), are both prime numbers. Going further to numbers 54 and 66, we find another symmetry emerging with 53 and 67 being primes and 55 and 65 not being primes; however, they are quasi primes. Numbers 47 and 49 to the left skip their matching pair and match up instead with the next pair of 77 and 79 (centered around 78 = 13×6), while 71 and 73, both prime, match up with 41 and 43 (centered around 42 = 7×6), also primes. Further out, we get the symmetry back with pairs 35 and 37 matching up with 83 and 85.

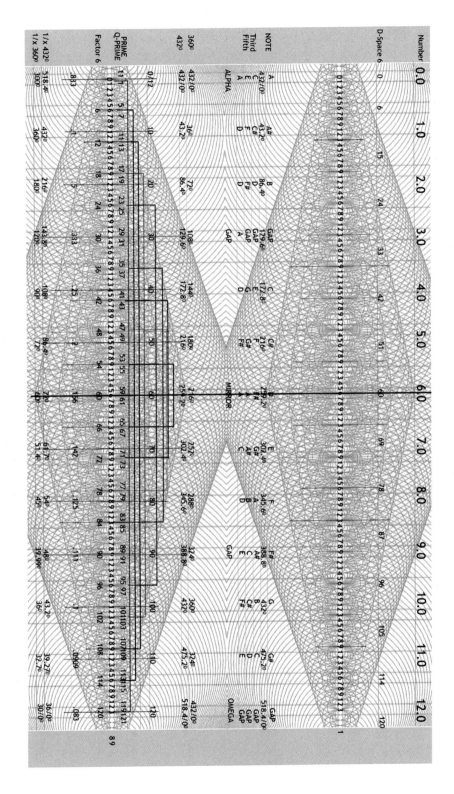

Figure 47: An illustration of the wave matrix generated from two-point-like interfering sources and for a scale between 0 and 121, along with prime and Quasi-prime symmetry, angular references, and musical notes correspondences. On the main horizontal axis of the wave interference matrix, we find the digital roots of numbers from 1 to 121. Listed on top of them are prime and quasi-prime numbers positioned in a symmetrical pattern around 60, as explained above. Above the level of prime numbers, we have numbers' angular references as a fraction of 360° and 432°. So, above 60, we have 0.6×360° = 216°, while 288° is just 0.8×360° and so on. We use the decimal fraction of the numbers to maintain the degree references within the 360° circle. Notice how the cycle ends at 120, corresponding to 432°, which is the same value for the Pythagorean tuning. The axis below the digital root numbers is the 1/x reference of the numbers along with their degree references. So, 1/6 = 0.166… and 0.166×432° = 72°, and so is the case for the rest of the numbers. The top rows are reserved for the corresponding 12 musical notes and their degrees of reference.

58

When we reach numbers 30 on the left and 90 on the right, we lose our reflection symmetry for the pairs [29, 31] and [89, 91]. This is because, at 90, we have reached the fractal limit of our 9-base system, starting with 9, then 90, then 900, etc. Therefore, the break in the symmetry here indicates the end of one cycle and the beginning of another.

One other cycle starts/ends at number 24, where the appearance of twin-primes on both sides of multiples of 6 ends, and the first quasi-prime number appears, number 25. At 114, we get another break of symmetry, paving the way for the appearance of the first pair of quasi-primes, [119, 121], with the cycle finishing at 120, as in a full octave.

Of course, the above argument about numbers' symmetry doesn't necessarily require wave interference to be observed; we could have simply laid down the first 121 digits on a straight line and observed the same pattern as well. What we are suggesting, though, is that wave interference pattern is in itself a numeric phenomenon, where the interfered waves form a discrete space around the source points such that specific points within this space hold certain properties unique to themselves, as well as correlated to other points, based on their reference numbers from the source. This discrete (or quantized if you wish) space is determined by many parameters and variables, including the medium in which the waves propagate.

Therefore, this wave dynamic of numbers suggests that the interference pattern of the waves creates a disturbance in space, quantizing it into fractals of regions that are not only described by wave physics but also by the numeric values of the nodes (where the wavefronts converge) and their mutual properties (primeness, quasi-primeness, etc.).

The above wave matrix extends to many physical phenomena, including electromagnetism and sound. Going back to the above figure, the upper part of the graph, where it says *Note*, illustrates the matching of the 12 musical octaves with the 12-based (2×6) fractal of the wave matrix, as if the interference is generated by sound waves. We used the notes of the major 5th octave with the A5 tone tuned to the Pythagorean standard pitch of 432 Hz because by using this tuning value, we get a perfect correspondence between the angular references of the notes and the internal angles of specific polygons, as will be explained in detail in Part III, among many other properties unique to this number.

Hence, the wave interference pattern is a unifying matrix where numbers, as well as geometry, sound, and everything that observes the wave phenomena, can be expressed. In return, it enables us to map the fields around these phenomena into discrete numeric points of specific connections and entanglements, creating a holistic matrix that brings all of these seemingly different disciplines into unity and harmony. Number 6 plays the center role or the mirror of this matrix, where numbers on both sides have some sort of connection or reflection to each other. Whether 6, 60 or 600, etc., everything seems to revolve around this amazing number.

The Wave Dynamic of the Fundamental Constants

Most of the fundamental laws of nature rest upon a set of constants that governs their behavior and, consequently, determines the universe's physical reality. Still, there is no well-understood theory capable of explaining why these specific numbers became so fundamental to nature. One famous example is the fine structure constant ($\alpha \approx 137$) described by the American physicist Richard P. Feynman as *"a magic number that comes to us with no understanding by man. You might say the 'hand of God' wrote that number, and 'we don't know how He pushed his pencil."* Thus, finding the origin and reasoning behind these constants will open the door for a much better understanding of the physical world and will enable science to cross boundaries never thought possible before.

In this section, we demonstrate how mathematical and physical constants emerge as combinations of numeric amplitudes and phases embedded within the above-explained wave dynamic of numbers and its fractal fabric. This novel approach will enable a better understanding of the origin and structure of these constants. Furthermore, it provides the means to discover new ones that have not been identified yet.

Almost all fundamental constants are irrational or transcendental, like π or e. Consequently, each constant can be written as a principle integer representing its particle-like aspect, an *amplitude*, plus a sequence of randomly fluctuating digits representing its wave-like aspect, a *phase*. So, a number like $\pi = 3.14159...$ can be decomposed into an amplitude of 3 and a phase of .14159..., as shown in the figure below.

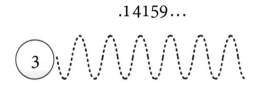

Figure 48: The number π, expressed in a wave/particle fashion as the sum of an integer amplitude (3) and a phase of 14159...

By the same token, the golden ratio $\Phi = 1.618...$ emerges from number 1 by adding a phase of $0.618...$ to it, which is also the value of its inverse: $\Phi = 1+1/\Phi$. The square root of 2 ($\sqrt{2}$) emerges from the same number 1 by the addition of a different phase, $0.4142...$

In theory, any number can be a potential amplitude for constants; we only need to attach a proper phase to transform it from its *static* form to a wave-like *dynamic* one.

Before we delve deeper into the wave matrix of constants, we first need to elaborate on two important points. The first point concerns magnitudes. We are used to considering numbers like 3.14, 0.134, 31.4, etc., to be very different from each other, especially number 3.14, which relates to π, a fundamental number used to calculate many mathematical and physical qualities, whereas the rest are not. The same thing can be said about any other number, especially those with unique mathematical properties such as e, Φ, etc.

While the decimal point is important to determine the magnitude of numbers, still, it is a *relative* quality. In other words, when we say that the speed of light $C = 3 \times 10^8$ [m/s], the 10^8 factor indicates how fast light is compared to a speed of 1 [m/s], for example, or to the speed of sound, being around 343 [m/s]. However, in this frame of units, the decimal point changes depending on the scale while the main number will not. So, for C again, if we were to use Km/s instead, the factor would reduce to 10^5 while the amplitude integer (3) would remain the same. Similarly, we can use 3.14... for π, or 10×.314... or 0.1×31.4..., etc.

In other words, fundamental constants are unique to the universe, mainly *for their numerical sequencing much more than for their relative magnitudes*. The decimal point only creates fractals of the same number. Therefore, we will handle numbers with some liberty in regard to their decimal point.

The second point concerns the numbers used to reference the various constants of nature. In general, fundamental constants have specific numeric references. For example, number 3.14159... is the familiar form of π. However, $1/\pi = 0.3183...$ is a different reference to the same constant, just like 0.31415..., 31.415..., etc., which are as valuable to us as the original form is. Similarly, $\Phi = 1.618$ and $1/\Phi = 0.618 = \varphi$ are both valid references to the same constant.

There are many other ways to reference constants depending on what mathematical operation we apply to them, such as squaring, square or cube roots, exponentials, etc. One important method is to convert the decimal reference of the constant into an angular one that positions it around a unitary circle. So, for a number like $\varphi = 0.618...$, we find its angular reference by multiplying it by 360°: 0.618...×360 = 222.48... This new reference of Φ (222.48°) is no less informative than its original value. This angular method is crucial in discovering the quadrant-relationships constants have with each other.

By explaining the above two points, we hope to remove any ambiguity that might arise from our constants assignments policy, as many of the constants we are going to encounter across the wave-matrix will appear as one value or the other; however, all can be traced back to their one familiar form.

Let's start exploring the wave pattern shown below. The pattern is made of 11 numbers, with the 12th initializing the beginning of another cycle or fractal, a higher octave of number 1. Thus, the pattern is centered around number 6, which works as a geometrical mirror across which numbers add up to 12: 1+11, 2+10, 3+9, etc. Similarly, each of the 12 integers that define the periodicity of the numerical wave works as a potential amplitude surrounded by a couple, or more, mathematical and physical constants that emanate from these amplitudes, as shown on the top part of the numerical matrix in the below figure.

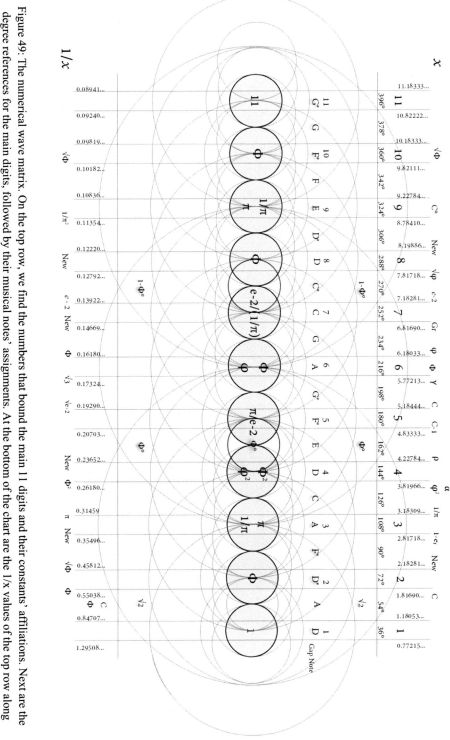

Figure 49: The numerical wave matrix. On the top row, we find the numbers that bound the main 11 digits and their constants' affiliations. Next are the degree references for the main digits, followed by their musical notes' assignments. At the bottom of the chart are the 1/x values of the top row along with their constants' affiliations.

And as every number has its own unique reciprocal value, the bottom part of the pattern is reserved for the inverse of the amplitude numbers and their corresponding potential constants. Thus for each number, there are three other entangled numbers, in accordance with the quadrant symmetry principle we talked about in chapter 2.

Starting with number 6, the mirror, the two surrounding constants are $\Phi-1 = \varphi = 0.618...$ (or 6.18..., as we agreed on the irrelevance of the decimal point. (Think of it as if we are working with numbers from 0.1 to 0.11 centered around 0.6 instead.) The reflection constant around 6 is Euler-Mascheroni constant $\gamma = 5.77...$ (formal value is 0.577...), which is a constant that recurs a lot in number theory, especially in the expansion of some functions. The inverse aspects of these two constants are Φ and 0.17324..., which is almost identical to $\sqrt{3} = 1.7320...$ And as explained earlier, number pairs on both sides of the mirror number 6 complete each other to 12, which also applies to the constants surrounding these numbers. So $\varphi + \gamma$ should equal 12 also, which is, in fact, the case: 6.18... + 5.77... = 12.

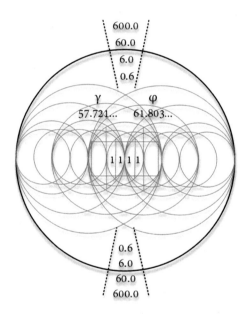

Figure 50: Numbers 60 or 600 or 0.6, etc., create the same mirroring effect number 6 does. For the case of 60, it will also be surrounded by fractals of $\Phi \times 10$ and $\gamma \times 10$, as indicated.

Moving to number 7, its two bounding constants are $e - 2 = 7.18...$ and $Gr = 6.8169...$, which is a new constant (called the Grant constant) that is believed to have a relationship with the dimension of time. The $1/x$ aspects of these numbers are 0.13922, and 0.14669, which are new constants whose functions have not yet been identified.

Number 5, on the other side of the mirror, is surrounded by constants that are the complementary pairs of those surrounding 7, being 12 – 6.8169… = 5.1844… and 12 – 7.1828... = 4.8333… These two constants represent the speed of light (in miles/sec) C and C-1, respectively. This is because 0.5184 × 360 = 186.624 (the speed of light in miles/sec), and 4.8333 is just 1 – 0.51844… Their inverse aspects are 0.19290… and 0.20703, with the first being $\sqrt[3]{[(e - 2)/100]}$.

Number 4 is bounded to its left by ρ = 4.227… = 10 - 5.7721... (a fractal of γ), and to the right, we have 3.819… = φ^2 ×10 with φ = 0.618… Their reciprocals are 0.2365…, which is a new constant, and 0.2618… being a fractal value of Φ^2. On the other side of the mirror, number 8 is bounded by 12 – 4.2278… = 7.8128…, close to $\sqrt{\varphi}$. The values of their corresponding inverses, 0.1279… and 0.1222…, are new constants. The table below lists the rest of the numbers and the constants that bound them, along with their identities, if known.

CONSTANTS		1	11	2	10	3	9
x	Left	1.180	11.183	2.1828	10.183	3.183	9.227
	Identity	New	New	New	New	$1/\pi$	New
x	Right	0.772	10.822	1.810	9.8211	2.817	8.784
	Identity	New	New	C	New	e_f	New
1/x	Left	0.847	0.089	0.4581	0.0981	0.3145	0.1083
	Identity	New	New	$\sqrt{\Phi}$	New	π	New
1/x	Right	1.295	0.092	0.5503	0.1018	0.3549	0.1135
	Identity	New	New	Φ	New	New	New

Table 12: Numbers [1, 2, 3, 9, 10, 11] and the constants to which they and their inverses correspond.

Each main digit of the periodicity has a musical tone assigned to it, illustrating the correlation between constants and music. The whole system, therefore, is a vibrating matrix of numbers, sounds, and geometry, manifesting into the physical reality via the constants and their phases and the musical tones and their frequencies.

Constants Transformation and Unity

As argued above, the many mathematical and physical constants are simple reflections or transformations of each other through some basic mathematical operations. As we have already seen, one such operation is related to the 12-based complementary property of the wave pattern, where the sum of each two constants paired around number 6 is equal to 12. Another transformation is the inverse operation, where numbers on the upper part of the wave pattern are inverted to generate those at the bottom part.

A third method originates from the quadrant symmetry that constants exhibit when transformed into angular references across a unit circle, explained in detail in chapter 2. By positioning a constant at a specific angle around a circle, three other angular references are automatically determined, as explained in the figure below.

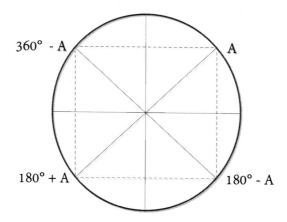

Figure 51: The definition of the quadrant angular complementary relationship between a set of four numbers.

For example, the angle of 261.8° corresponds to $\Phi^2 = 2.618$. The complementary value of this angle across the north/east axis is 98.2°. By normalizing these values with respect to the 360° of a full circle, their sum will equal to 1: 261.8/360 = 0.72722... and 98.2/360 = 0.27277... What is interesting is that the value of 0.272 is nothing but $\sqrt{\Phi}$ - 1. Therefore, we have Φ^2 on the left quadrant, while on the right quadrant, it is mainly $\sqrt{\Phi}$. The quadrant complementary value of 261.8° across the east/west axis is 278.2°. This angular reference is related to π through the following relation: $(1 + 0.77277...)^2 = \pi$.

The final angle of this quadrant set is 81.8° with 81.8/360 = 0.22722... Now, 0.22722...×432 = 98.16, and 98.16/360 ≈ 0.273 ≈ 4/π -1. This quadrant set, along with its corresponding angular and constants references, is shown in the figure below.

The angular reference of 432° is very important for detecting the various numbers/constants relationships, just like the 360° based one is. And as we will discover later in the book, the frequency value of 432 Hz is crucial in establishing the connection between music and geometry as well as in bringing the musical octaves back into their natural doubling order.

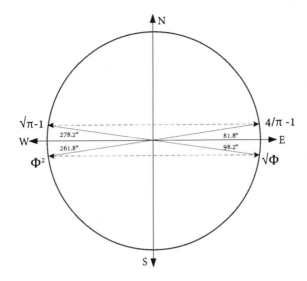

Figure 52: One quadrant set corresponding to the two constants π and Φ and their mirror images.

The table below illustrates how some of the most fundamental constants can be derived from each other by using very simple mathematical operations.

Constant	Symbol	Calculation	Value
Light Speed	C (mile/sec)	$(\frac{e}{\pi} + 1) \times 10^5$	186.525
Euler	e	$9 - 2 \times \pi$	2.71681
Fine Structure	α	$137.5 - \sqrt{(\frac{\varphi \times 360}{1000})}$	137.036
Pi	π	$\Phi^2 \times \frac{432}{360}$	3.14150
Euler-Mascheroni	γ	$\frac{\left(\frac{\Phi}{10} \times 360\right) - (1 - \sqrt{(\frac{\varphi \times 360}{1000})})}{10^2}$	0.57721
1/Phi	φ	$\frac{\left(\frac{e-1}{10} \times 360\right) - \frac{\sqrt{C}(\frac{km}{s})}{10^4}}{10^2}$	0.61803
Planck Length	pl	$\frac{\sqrt{10} + 3}{10} + 1$	1.61622

Table 13: The various transformations of the most fundamental constants of nature.

Many of the transformations in the above table involve the golden ratio Φ or φ = Φ - 1. Even the golden angle of 137.5° can generate many fundamental constants by the simple arithmetical transformations encountered thus far. The golden angle is defined as the angle of an arc that sections a circle into two parts such that their lengths satisfy the golden ratio rule (their ratio equal to Φ), as shown below.

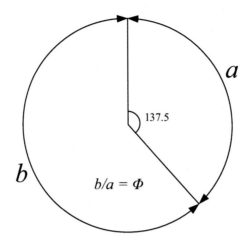

Figure 53: The golden angle defines an arc such that it sections the circle into two parts, having a ratio of Φ.

We initiate the golden angle based transformations by taking its ratio in respect to the angular references of 360°: 137.5/360 = 0.31894..., and 1/0.38164 = 2.61818..., which is the square of the golden ratio (Φ^2). Also, 360 - 137..5 = 222.5, which is the angular reference of Φ. It is also the quadrant reflection of 137.5 with respect to the vertical axis. (The remaining two members of this quadrant set are 42.5° and 317.5°.) Additionally, 137.5 - √0.2225 = 137.03... = α, the fine structure constant.

In respect to the 432° angular reference, we find that 137.5/432 = 0.31828... ≈ 1/π. Now 0.38194... ×432 = 165   and 432/360 ×165 = 198. These two numbers, 198 and 165, are very related, as their ratios are 198/165 = 1.2 = 6/5 and 198/6 = 165/5 = 33. This hints at a hexagonal/pentagonal relationship between the various numbers of this transformation.

The hexagon and the pentagon establish one of the most fundamental geometric relationships ever. Together they are referred to as the Hexapentakis, to which we dedicate a complete chapter in in an attempt to fathom its secrets.

In conclusion, it would not be far-fetched to speculate that, in reality, there is only one fundamental constant, with all the others being simple transformations of it. These transformations rest upon the various symmetries of the numerical wave matrix, along with the quadrant angular references of these constants and their 1/x aspects.

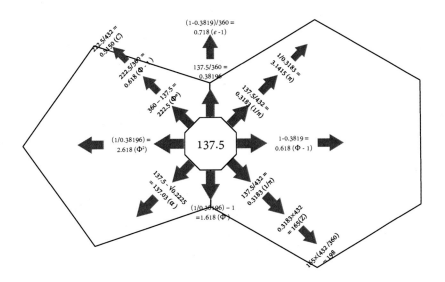

Figure 54: The transformation diagram of the golden angle (137.5°) through the 432 (hexagon)/360 (pentagon) angular references.

One perfect candidate for this quintessential constant is the golden ratio (Φ) and its inverse (φ) (as well as its angular reference 137.5°), as they all reside at the core of the wave matrix, emanating from the central mirror of number 6. As we have seen earlier, this is a constant that controls so much in nature, from the geometry of the farthest galaxies to the proportions and growth-rates of all living things, to the magical quartet, etc. It definitely deserves to be referred to as the *primordial constant*, the first one to be generated, and the origin of all other constants.

> *"Numbers are the nouns, constants (irrational) are the verbs, and geometry*
> *organizes the syntax of the universal language of consciousness."*
>
> - Robert Edward Grant

The Reciprocity of Numbers and

Prime Factorization

"The obvious mathematical breakthrough would be the development

of an easy way to factor large prime numbers."

-Bill Gates

The Multiplicative Inverse

The reciprocal of numbers, also known as the *multiplicative inverse*, is one field of number theory often studied with modular math. It is defined by the remainder it generates, which is the digits resulting from dividing number 1, the numerator, by a number (*x*), the denominator. And though it is one of the simplest and most used operations in mathematics, it may be one of the least studied and understood. As we discussed earlier when we talked about the wave nature of numbers, the remainder varies depending on the denominator (*x*), where it comes in one of these three different flavors (considering that we are only inverting rational digits, not irrational or transcendental numbers):

- Rational numbers, e.g., 1/2 = 0.5.

- Repeated same decimals, e.g., 1/3 = 0.333...

- Repeated sequence of numbers (period), e.g., 1/7 = 0.142857 ...

Taking the reciprocal of the first nine digits reveals some interesting relationships between them. Looking at the table below, we see that numbers 2 and 5 are unique in the sense that their reciprocals exhibit a mirroring behavior, with number 2 inversed to 5 (0.5) and vice versa for number 5 (0.2). Numbers 3 and 6 produce repeated periods of themselves, threes and sixes, respectively (similar to doubling). The reciprocal of number 8 produces a digital root of itself $D(125) = 8$, while that of number 9 brings back the period to unity (number 1).

x	1	2	3	4	5	6	7	8	9
1/x	1	0.5	0.333...	0.25	0.2	0.1666...	0.142857	0.125	0.111...

Table 14: The multiplicative inverse of numbers from 1 to 9.

We argued earlier that those numbers whose reciprocals produce infinite periods, such as 1/7, are basically numerical waves that exhibit fundamental wave-like properties, as in their forms and interference. In this chapter, we will concentrate mainly on the periods of prime and semiprime numbers as they are sure to produce such waves.

Similar to the main three categories of numbers' reciprocals, prime numbers' reciprocals also come in a variety of three flavors, depending on the length of their period:

- Full period primes, being equal to the prime numbers minus 1, as for numbers like 7 (period of 6) and 19 (period of 18).

- Half period, where the period is equal to the prime number – 1, divided by 2, like numbers 31 whose period is (31-1)/2 = 15.

- Neither full nor half, like number 137, whose period is eight digits only (notice that (137-1)/8 = 17, another prime number).

Most prime numbers produce perfect numerical waves, especially those having even periods (full-period), as one half of the period is always a completer to the other half, adding up to a series of nines. This is contrary to the reciprocal of prime numbers, whose period is odd (half period).

The perfect way to visualize this property is through the geometry of the circle; by distributing the individual digits of a single period around a circle, the completion pattern becomes obvious through the relationship between diagonal digits, as shown below for 1/19. This completion property is not restricted to the reciprocal operation only; it is the hallmark for most numeric sequences under the digital root operation, as we saw in Fibonacci, Lucas, etc.

The reciprocal operation is unique as it is probably the only mathematical operation where applying it twice to a certain number returns the same number back again. It works like a mathematical mirror, reflecting back and forth between the number and its inverse, ad infinitum.

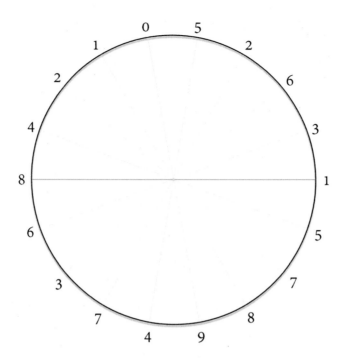

Figure 55: Distributing the 18-based period of the reciprocal of number 19 around a circle reveals the complementary property of its individual digits, with every two digits sharing the same diameter adding up to 9.

Some numbers have unique relationships with their inverses. The best example is the golden ratio Φ = 1.618... with its reciprocal value returning itself minus 1: $1/\Phi = \Phi - 1 = \varphi = 0.618...$ Another unique example is the square root of 10. This is because its inverse returns the same square-rooted number, only one order of magnitude smaller. $\sqrt{10} = 3.162277...$ and $1/3.162277 = 0.3162277...$

Due to their polar nature, numbers and their inverses form two opposing and complementary wavefronts, like sine/cosine waves, as shown below. These two sinusoidal waves get closer and closer to each other until they cross at a point where the two values of x and $1/x$ become identical, as in the case of $\sqrt{10}$ and its inverse (identical in their numerical sequencing, not in their relative magnitude). The crossing point creates a singularity node where x and $1/x$ flip and continue along the wave pattern until the next node, which happens at $\sqrt{100}$ and its inverse (not shown), where they flip again and so on, very similar to the helical form of a DNA strand. The nodes get further apart as the ratio between the singularity number and its inverse increases from 10 to 100 to 1000, and so on.

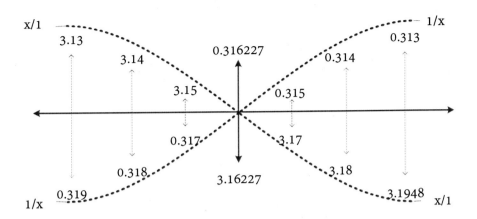

Figure 56: Two wavefronts made by numbers and their inverses around the vicinity of number 3. The two waves intersect at the node of 3.16227... where x and 1/x match up, except for their decimal point.

For some numbers, the remainder of their reciprocal exhibit symmetrical properties with its multiples. Let us consider the pattern of $n/31$, for n ranging from 1 to 31. Number 31 is a half-period prime, with its reciprocal repeating in a period of 15 digits. As shown below, the first thing we notice is how the 15th digit of each period follows a cascading pattern from 9 to 0 to 9, and so on for a total of three complete cycles. There are also horizontal mirror relationships in the digital root of positions: the horizontal symmetry adds to 13 all the way to the top of the circle. Interestingly, the 14th digits repeat in the triple groups of [1, 4, 7], [2, 5, 8], and [3, 6, 9].

The above properties of the reciprocals' periods are clear indications that these trailing digits are not as random and trivial as previously believed. There are so many hidden laws and relationships that can be extracted from their sequencing, especially in regard to the nine-completion property, which we explored in more detail in chapter 4.

Prime numbers hold a unique status regarding this operation, not only because they produce some of the most perfect periods, but also as we will discover next, because hidden within the periods of their products' reciprocals lies far more interesting and fundamental information than ever thought before.

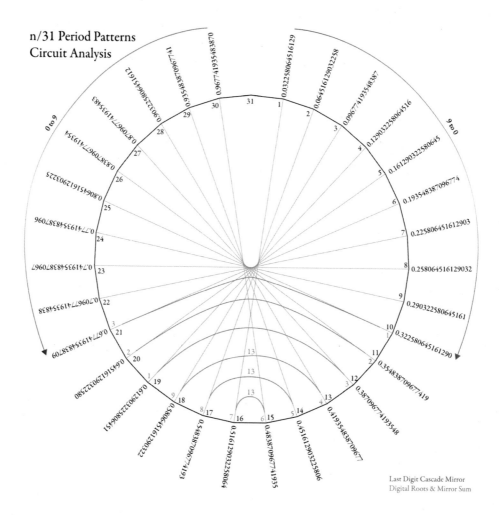

n/31 Period Patterns
Circuit Analysis

Last Digit Cascade Mirror
Digital Roots & Mirror Sum

Figure 57: The circular analysis of the *n*/31 case. The last digit (15th) of the decimal period repeats in descending order from 9, 8, 7…2, 1, 0, 9, 8…, etc. There is a mirror relationship around the horizontal axis with numbers adding to 13 all the way to the top of the circle (only the lower six levels are shown).

73

Prime Factors in the Decimal Period

We have already talked about the semiprime factorization problem and its paramount importance for science and technology, such as encryption of data and personal privacy. We also showed how we could use the 24-wheel of numbers, along with quasi-primes and digital root math, to speed up the factorization process. (As a reminder, quasi-primes are numbers produced from the multiplication of a couple of primes, excluding numbers 2 and 3.)

But how about inverting semiprime numbers; what information can we find in their remainders? As the reciprocal of a semiprime is the product of the reciprocals of its two prime factors, could the remainder inherit any information about its prime factors or their reciprocals? This is a very profound question in the sense we are comparing the two primes to two strands of DNA, of a mother and a father, coming together to form a baby semiprime. In the biological sense, the fetus carries information from both parents, passed on within its DNA. Could the reciprocal of the semiprime be carrying similar information or hints about its prime parents? Our investigation indicates that the hints are definitely there, no less than the prime factors themselves.

So, we would like to search for any trail of the parent primes within the mantissas (the trailing digits of the reciprocal) of the child semiprimes. The lengths of semiprimes range from 2 digits, like 10 (5×2), all the way to any number we can think of. Their reciprocals' periods can be even much longer. Therefore, we start with the simple case of 12-bit semiprimes, which translates to a 4-digit long number (two 2-digit-long primes multiplied together). We consider the simple case of $A = 31 \times 83 = 2573$. As the square root of 2573 has 2-digit order of magnitude (50.724…), we look for a couple of numbers having a similar magnitude while satisfying the search criterion we talked about earlier, such as the last digits criteria, the digital root criteria, etc. For our case, as the last digit of the semiprime is 3, the only options for the last digits of the two factors are either [1, 3] or [7, 9]. The digital root of the semiprime is $D(2573) = 8$; therefore, the two primes' digital roots should be either [1, 8] or [2, 4]. The full reciprocal of the inverse of 2573 is shown below (which has been captured from Wolframalpha.com). Interestingly, we were able to find the two parent-primes hidden inside the period, as shown in the same figure, highlighted in yellow.

After succeeding in finding the prime factors (one or both) for almost all 12-bit semiprimes, we randomly tested 4,027, 24-bit public keys provided by the RSA official website. This translates into prime factors of 4 digits long each. The incidence of both prime factors appearing within the semiprime's reciprocal exceeded 95%, with 98% of the sample demonstrating the incidence of at least one prime factor within the semiprime reciprocal value.

0.0003886513797123979790128254955305091333074232413525068

013991449669646327244461717839098328790672366886902448

503692188107267780800621842207539836766420520792848814613

291877186164010882238631947143412359113874854255732607850

757870190439176059075009716284492809949475320637388262

728332685581033812670034978624174115818111115429459774582

199766809172172561212592304702681694520015546055188495919

160513019821220365332296929654100272055965798678585308977

846871356393315196268946754760979401476875242907112320

248736883015934706568208317139525845316750874465604352895

452778857364943645549941702293043140303148076175670423

3

Figure 58: The two prime factors of the semiprime 2,573, 31 and 83, can be found embedded within its reciprocal decimal extension (highlighted). Image captured from www.Wolframalpha.com.

We also searched for the palindrome of the prime numbers, as in these numerical waves, as well as in regular ones, there is a reflection symmetry in both directions of the wave. The statistics of our finds are listed in the table below.

Found p1 and p2 and their palindromes	3765
Found p1 and p2	3842
Found only p1	54
Found only p2	58
Found p1 and p2 palindrome	3826
Found only p1 palindrome	68
Found only p2 palindrome	67
Missed finding primes	73
Missed finding palindrome primes	66
Found none	39

Table 15: Statistical analysis of the number of times prime factors, p1 and p2, and their palindromes have been found within the reciprocals' decimal extensions of a sample of 4027 semiprimes, each being 24-bits long.

From the table shown below, we find that the reciprocal's decimal extension's mean incidence is 74, for both the first and second prime factors. It is also 74 for their respective palindromes. This is a very interesting equilibrium of incidences that confirms the numeric symmetry of the waveforms, which deserves more research and verification.

Average Decimal Repetition Period Length	2474730
Average p1 average frequency	74
Average p2 average frequency	74
Average p1 palindrome average frequency	74
Average p2 palindrome average frequency	74
Average distance from begin of period to first p1	9349
Average distance from begin of period to last p1	735427
Average distance from begin of period to first p2	9630
Average distance from begin of period to last p2	735350
Average distance from begin of period to first p1 palindrome	9342
Average distance from begin of period to first p2 palindrome	735680
Average distance from begin of period to last p2 palindrome	9084
Average distance from begin of period to first p2 palindrome	735360

Table 16: Statistical analysis of the average frequency of the appearances of p1, p2, and their palindromes, along with their average distances from the beginning of the repetition period of the decimal extension.

To test the statistical significance of the above results, we conducted a Z-test on our sample. Our Z-value was calculated in the range of -11.65. Therefore, the result we are seeing cannot be attributed to mere statistical noise or randomness.

In its simple form, this method is not very reliable for factorizing larger semiprime numbers, as calculating the semiprime inversions' mantissas consumes a large amount of calculation power. Nevertheless, it could be one of the fastest and most accurate factorization methods when combined with some mathematical tricks. One of these mathematical tricks is explained in the next chapter.

Still, no matter how difficult this method for factorization is, the fact that the prime factors exist within the semiprime mantissa is a breakthrough that will definitely challenge our perspective about prime numbers and their products as well as the whole multiplicative inverse operation.

The Decimal Jumping Method

For 24-bit size prime numbers, the complexity of finding the prime factors within the reciprocal period can be considered moderate. However, for larger keys, the complexity increases due to three main factors:

1.   The search space becomes much bigger.

2.   The prime factors increase in length as well, which means the number of consecutive digits to be tested is larger and hence requires greater processing power and time.

3.   Converting the period into a string of digits for indexing and searching is also a very time-consuming process.

Consequently, testing larger bit-lengths, such as 256 and 512 bits, has proven to be a very lengthy process. For example, we were able to find one prime factor for a (96-bit) semiprime number; however, it took around 60 days to complete. But still, it was there.

One optimum solution to this problem is to be able to predict the range for where the prime factors may be located within the mantissa and then jump to this location without the need to string-convert the whole decimal period, which could be hundreds of thousands or even millions of digits long.

Unfortunately, predicting the location of the primes has proven to be a difficult task; as for now, it seems there is no apparent order for their whereabout within the period. Jumping within the period, however, seems to be a much easier task to accomplish, as we were able to find a method for consistently rotating the reciprocal period such that the number pattern starts in a new location of the same period. We are calling this process the *decimal jump*.

Mathematically, the decimal jump is achieved by multiplying the reciprocal of the semiprime by the factor (2^n), which is a binary expansion to the order n. For each semiprime, there is a sequence of exponents n that perform successful jumps to different indices within the period, while for other exponents, the jumps land on different decimal periods entirely.

Let us take semi-prime 5767, for example, whose prime factors are 73 and 79, and whose reciprocal repeating decimal period is 104 digits long:

[17340038148083925784636726200797641754811860586093289405236691520721345586960291 3126 40887809953181897000].

By applying a binary expansion of an exponent $n = 9$, we get the same period back, but starting 85 digits later, around 82% of the whole period: $2^9 \times (1/5767) =$

[08878099531818970001734003814808392578463672620079764175481186058609328940523669 1520 72134558696029131264].

For an exponent of $n = 18$, we jumped over half-way through the sequence:

[45586960291312640887809953181897000173400381480839257846367262007976417548118605 8609 32894052366915207213]

Up till now, the decimal jump has not been a predictable process. In most tests, a jump can initiate the number sequence from nearly anywhere in the period; it is just as simple to start at 80% of the way through the period as it is at 50% or 10%, etc. However, we do believe that there must be a way to predict the range; we just need more research to be able to understand and connect all the numeric variables that affect the logic of the problem.

One unexpected find of this study is observed when we plot the sum of the frequencies of appearance of both prime numbers versus the period length of the semiprime reciprocal. We found that the points line up along a couple of straight lines, as shown below. Thus, there seems to be a direct linear relationship between the period of the semiprime and the number of times its two prime factors show up within the period. This linear behavior has been observed for 24-bit, 28-bit, and 32-bit semiprimes. Interestingly, the steeper line (grayish) belongs to the 28-bit and 32-bit public keys, while the darker line belongs to the 24-bit public keys only.

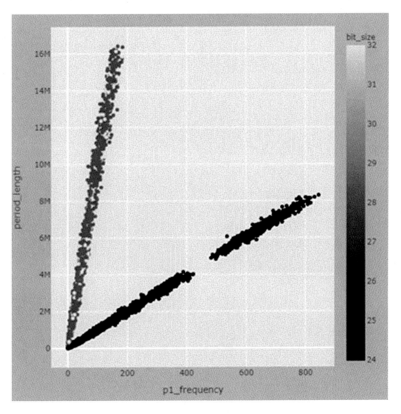

Figure 59: A plot for the frequency of incidence of the two prime factors, for 24-bit (dark), 28- and 36- bit (gray) semiprimes, vs. the period length of the semiprimes. Notice how the points fall along a couple of straight lines, exhibiting a strong linear relationship that is dependent on the size of the semiprimes.

If this observation is scalable to all bit-sizes of the semiprime, and we have the reasons to believe it is, it allows us to calculate the probability of finding the prime factors within the period of any semiprime having any bit size by calculating the slope from already known semiprimes and their factors of a certain bit-size. This could help us also estimate the whereabouts of the prime factors within the period, consequently leading to very fast factorization. This observation is another strong indicator of the presence of the prime factors within the mantissa of their product.

The Reciprocal Operation:
Interpretation and Meaning

All the above points toward verifying our original hypothesis: that a semiprime reciprocal carries information about its prime parents, just like a DNA strand does. This hypothesis opens the door for a complete re-understanding of this operation and its importance. To achieve this goal, however, we first need to understand the meaning behind the multiplicative inverse's mathematical logic.

When we take the inverse of a number x, what we are really doing is dividing unity (number 1) into x parts. We are kind of normalizing the number, comparing its value to 1, which stands now as the utmost upper limit. The larger the number, the smaller its inverse will be. Hence, we are reversing the numeric logic, where the largest becomes the smallest, and vice versa.

In the original number space, x has no upper limit to compare to, except for infinity, which is, by definition, undefined. On the other hand, the inverse operation creates this bounded domain, from 0 to 1, where all numbers can exist, no matter how large or small they are. (Some may argue that 0 itself is not defined, which renders the lower limit of this domain undefined). This, by itself, is an amazing concept, where the whole vast infinite space of numbers can be compressed and encapsulated within the limits of 0 and 1, by a simple operation, where every number has its own pair, or mirror, on the other side of the boundary. Within this bounded space, numbers can be literally infinite (having infinite mantissas), while outside it, where the space is unbounded, numbers are finite and well behaved.

At the boundary, between the two domains, resides number 1, the one and only number that belong to both domains simultaneously. It is its own inversion, its own mirror, the gatekeeper; for any number to cross the boundary, it needs to undergo the reciprocal process. In a sense, it is like the boundary between the macrocosmic and microcosmic worlds, or between the classical and quantum domains, where inside the boundary, the world is very different from outside it, but it is still highly entangled with it.

From another mathematical perspective, the multiplicative inverse of a number x is an operation that produces a number when added to itself x times, returns back to unity. For example, adding 0.125 (1/8) to itself eight times gets us back to number 1. So, each number in the unbounded domain has its own unique fraction of unity in the bounded domain. Both are highly entangled, and forever bonded by the relationship: $x \times x^{-1} = 1$. Just like the rainbow, where adding the different colors produces white light, the number 1, and from this same number comes the different colors through refraction, the inverse.

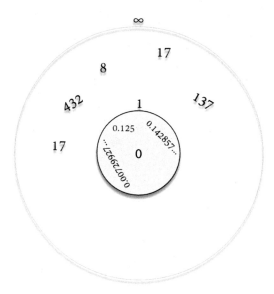

Figure 60: The reciprocal operation creates a boundary between two numeric worlds; one is unbounded, with no upper limit, from 1 to infinity, while the other is bounded, from 0 to 1. Both worlds incorporate all numbers, with every number having its own mirror on the other side. Only number 1 resides in both worlds.

From a physical perspective, the 1/*x* operation can be compared to the double-slit experiment, where a beam of particles at the slit (e.g., photons, electrons, etc.) transform into propagating waves, interfering with each other and creating the diffraction pattern we saw earlier in chapter 4. At the slit, each particle/ number undergoes a transformation that changes its essence from a point-like nature, such as 7, into a wave-like nature, such as 1/7 = 142857..., as shown below.

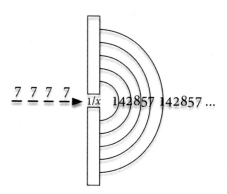

Figure 61: The 1/*x* operation *rese*mbles the slit experiment where particles of light or matter undergo a wave-like transformation of their essence, just like numbers do.

When we take the reciprocal of a number, we are peering deep inside it, dissecting it into its smallest constituents. By analyzing the remainder, we are analyzing its spectral and history, which allows us to deduce so much about its nature, as well as its origin.

The remainder is the trail the number leaves behind once it crosses the boundary between the infinite and the finite, between the rational and the irrational. And like any good trackers, analyzing the trail always leads to the source. And in this case, all trails lead to one.

Numbers Classification

"Numbers are the Highest Degree of Knowledge. It is Knowledge Itself."

-Plato

The concept of categorization is an essential tool implemented in most natural sciences. When things are put in their correct category, their order reveals so much about their identity and origin and their function and usefulness. For example, categorization helps scientists connect species together; by studying their resemblances and subtle differences, they can discern so much information about their natural habitat, diet, what led to this mutation in their forms or that change in behavior, etc.

And numbers are no exception. On the contrary, categorizing numbers is a powerful tool that enables us to reveal so much about their essence, their similarities and differences, their mutual relationships, etc.

We have already touched base on some different types of number categorizations, such as prime and composite, odd and even, rational and irrational, etc. We also talked about the different types of remainders the reciprocal operation generates. In this section, we expand much more on this specific type of categorization because it reveals so much about the nature of numbers and because it is intrinsically related to the other types of categorizations.

We start by examining the main categorization of numbers, not as a linear list, but in a circular fashion, as shown below, in what we call the Wheel of Categorization.

The first three main types of numbers are related to their polarity, odd or even. As you may remember, what determines a number's polarity is whether it is divisible by number 2, without any remainder left, or not. In other words, whether it can be expressed as $2 \times m$ (even) or $2 \times m + 1$ (odd), where m is some integer. However, if the number itself is infinite, as in irrational and transcendental numbers, we can't express it in either way; therefore, its polarity is undetermined (similar to its digital root). For that reason, the Wheel of Categorization is divided into three quadrants, Odd, Even, and Undetermined.

The next main subcategory is whether a number is a Prime, Composite, Irrational, or Transcendental. The Undetermined category already spans the Irrational and Transcendental subcategories. As all prime numbers are odd, with the exception of number 2, the Prime category belongs mainly to the Odd section of the circle.

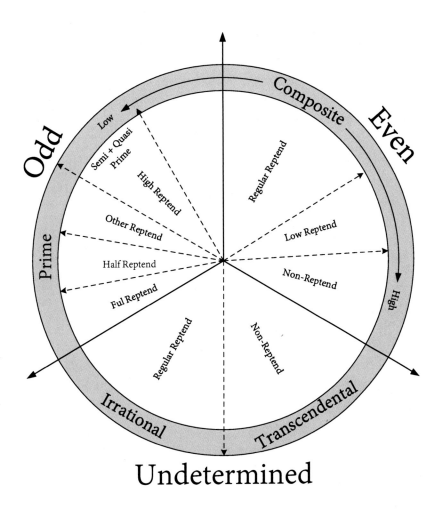

Figure 62: The Wheel of Categorization, where numbers are categorized based on their polarity, primality, rationality, and reciprocity.

The closest composite numbers to being primes are semiprime and quasi-prime. Those can only be odd also; consequently, they are right next to prime numbers in the Composite section of the Odd category. These numbers are known to have a low number of factors. On the other extreme, we find very high composite numbers, being always even. Therefore, they are positioned on the far end of the Even Composite subcategory.

It is worth mentioning something about numbers 2 and 3. Symbolically, these two numbers represent the first principles of polarity, odd and even, male and female, emanating from the eternal 1. Even though they are usually considered the first two prime numbers, many mathematicians do not include them in the prime category. This is because they both have intrinsic properties that set them apart from all other primes. Number 2 is the essence of evenness, as any number to be considered even must be divisible by 2. However, no other prime besides 2 is ever even. As for number 3, no other prime besides it has a digital root of 3 (and 6 and 9 as well.) And remember, when we distributed the numbers around the 24-wheel, 2 and 3 didn't belong to the prime moduli. That is why our quasi prime numbers started with the number 5 for their factors. Thus these two numbers are an anomaly when compared to all other prime numbers, and they definitely deserve to be isolated in a league of their own.

The next subcategory is about the reptend property of the reciprocals of these numbers. *The reptend* is a term used mainly to describe the repeating decimal or period of prime numbers. However, in our categorization wheel, we extended it to describe the period of all numbers.

The reptend property compares the length of the period of the number to its size. For example, full-reptend primes are considered the highest reptend numbers because their periods are equal to their value, minus 1. So, for a number like 19, its period is 19-1 = 18, as we have explained previously. And for these numbers, the period is always even.

But not all primes are full-reptend. There are half-reptend primes, whose period is $(p - 1)/2$ and is always odd, such as number 31, whose period is 15. And there are what we call here, other-reptend, as they do not follow a rule based on the prime size, like number 137, whose period is 8. The three categories are shown on the wheel, with the full-reptend on the far left of the odd category, followed by the half-reptend and then other-reptend.

In contrast to prime numbers, numbers with the most divisors (highly composite) possess the shortest periods. For example, number 60 has a period of one digit only, 1/60 = 0.01666..., while possessing the highest number of divisors for its size. On the far extreme, binary numbers (2^n), such as 1, 2, 4, 8, 16, 32, etc., possess no periods at all. Likewise, mathematical constants/irrationals/square roots, etc., possess no repeating period either, and therefore they are non-reptend.

Another important categorization we can deduce from the reciprocal property of numbers is related to how they behave on both sides of the mirror, the boundary between x and $1/x$. For example, some numbers on one side of the mirror are discrete or finite, with no trailing digits, while on the other side, they are continuous or infinite, like the reciprocals of prime numbers. Some are finite on both sides or are infinite, etc. (Note that we are considering any infinite trailing of digits for the remainder, not only the period or reptend type, as infinite, e.g., 1/3 = 0.333333... is also considered an infinite remainder.)

The table below lists the four types of numbers based on this categorization, along with some numeric examples. (Notice that the Infinite-Finite numbers are simply the reciprocals of the Finite-Infinite ones.)

	Finite-Finite	Finite-Infinite	Infinite-Infinite
Type of Numbers	Binary numbers 2^n	Prime numbers (except for 2, and 5), Trinary numbers 3^n, all multiples of 7, etc.	Irrational and transcendental constants
Examples	1, 2, 4, 8, 16…	3, 6, 7, 9, 11, 12, 13, 15, … 21, …	π, Φ, γ, e, $\sqrt{2}$, $\sqrt{3}$, etc.
Physical Phenomenon	May be related to time epochs and their derivatives, as in 60 min, 24 hours, etc.	Wave-Particle Duality	Controlling the physical reality: geometry, electron/photon interaction, radioactive decay, etc.

Table 17: Categorizing numbers based on the length of their decimal extensions after undergoing the reciprocal operation.

Among the first nine numbers, number 7 stands out as being the first prime to generate a reptend remainder for its reciprocal. It is the number that corresponds to the wave-particle duality, while at the same time defines some of the most important constants of nature, e.g., $\pi = 22/7$ and $e = 19/7$. Also, some binary numbers have a digital root of 7, as in number 16, for example, or number 1024, etc. In this sense, number 7 can intermediate between all the different categories of numbers. It is a wild card, able to cross boundaries and to enact the role of all other numbers.

No wonder then, number 7 is considered a divine number, the number of marriages between the material body (4) and the eternal soul (3), just like it is able to marry between all the above categories. It is also the number of cycles, especially times, with the days of the week forever locked in their septenary cycle. It is the number of the primary colors, as well as 7 the primary planets, the 7 heavens, the 7 earths, the 7 virtues, the 7 internal organs, the 7 sages of Greece, the 7 Rishis of India, the 7 chakras (energy points), etc. Even the Great Pyramid of Giza, some argue, is built upon the geometrical representation of number 7, the heptagon.

It is no wonder when all the numeric properties of this number tally perfectly with the symbolic meanings attributed to it over the millennia. It is the same ancient wisdom, rediscovered.

Number 6 sits between the prime numbers of 5 and 7. On its left, we find the well-behaved number 5, finite even after inversion, $1/5 = 0.2$. To its right, we find number 7, the first number to produce the infinite wave-like remainder, $1/7 = 0.142857…$ Number 6 has a reciprocal that repeats itself to infinity but without any period or wave-like pattern ($1/6 = 0.1666...$) Hence, number 6 kind of belongs to both realms; it is the boundary between the finite particle aspect (from 1 to 5, except for 3) and the infinite wave aspect (from 7 to ∞). So, number 6 is not only the central number for prime generation or the wave matrix fractal structure but also for the particle/wave-like nature of all number space.

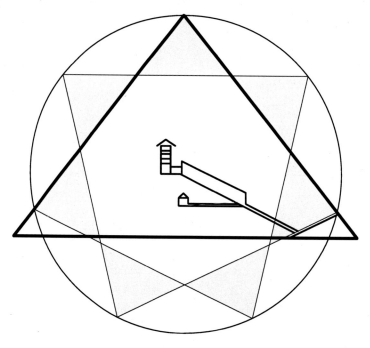

Figure 63: The heptagonal-based design of the Great Pyramid of Giza.

And with this, we conclude the first part of the book, hoping we have enriched, more than boggled, the mind of the reader, putting it in the right state, ready and prepared for understanding the rest of the discoveries of this book.

Part II

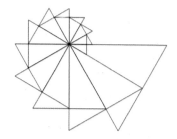

GEOMETRY AND SYMMETRY

"Let No Man Ignorant of Geometry Enter Here."

-Plato

Arithmetic and geometry can be considered the oldest branches of mathematics, if not of all science. Arithmetic concerns itself with the study of numbers and their properties, while geometry concentrates on spatial entities along with their qualities, such as shapes, distances, sizes, symmetries, etc.

The word *geometry* is of Greek origin where it means *Earth's measurements*. The word *measurement* in itself invites the concept of numbers, such as length, width, height, etc. Therefore, one is enticed to ask the question: which came first, numbers or geometry?

This is a tricky one, similar to that of the chicken and the egg, but on a more fundamental level (or is it?)

At face value, one argues that forms definitely came first, existed since the beginning of the universe. The fabric of space in itself, whether one-, two-, or three- dimensional, etc., is geometric in nature. And nature, in its essence, is random and often asymmetric. Observing random shapes, like trees, mountains, lakes, etc., doesn't readily invite the presence of numbers. However, when we delve deep into the minute constituents of matter, like the elementary particles, we find them following strict rules governed by numbers, like quantum numbers, the fine structure constant, the speed of light, π, e, to mention a few. Even the forces between these particles are defined by numerical constants, such as the Coulomb constant q, gravitational constant G, etc. In other words, for those minute particles to construct matter, they need the concept of numbers, one way or the other. Consequently, we may postulate that space is mainly based on geometry, while matter and forces are based on numbers.

Still, did one come before the other?

Probably there will be no definite answer, as they seem to have always come together. Therefore, one cannot discuss geometry without involving numbers, one way or the other, consciously or subconsciously. And the concept of numbers, by itself, is very abstract; it only becomes tangible when illustrated by geometry and forms. Moreover, there are so many mathematical problems that can only be solved through geometrical means, or with the geometric method offering the most elegant solution, as we will show later on for one such fundamental problem. What is important to realize is that for us to properly understand one or the other, we need to study them together, as an inseparable entity, similar to what we already have done in the previous part, like in the D-circles analysis or the prime numbers 24-wheel distribution, etc. Only together will their true quintessence reveal itself, and their secrets be uncovered.

Geometry: Forms and Categorization

"Geometry will draw the soul toward truth and create the spirit of philosophy."

-Plato

Classification through Numbers

We initialize our entry into the field of geometry by tackling the subject through the lens of categorization, which we did for numbers. There, however, we did it at the end of the study instead of the beginning, probably because numbers are a fairly abstract concept, and we had to build up some essential knowledge before we were able to properly sort them out. On the other hand, shapes are visual entities that entice you right away to put them into specific classification, based on their dimensionality, regularities, symmetries, etc. It is a subconscious reflection that requires no real motivation nor deep thinking from the observer.

In general, categorizing geometrical shapes is mainly based on numbers, like the number of edges, dimensionality, the sum of angles, etc. Still, other categories are not directly based on them, as for some types of symmetry, e.g., reflection or mirror symmetry. We will concentrate mainly on the cases where numbers and geometry are intertwined to gather as much information and understanding of both subjects as possible.

The simplest way to classify geometrical shapes is through their dimensionality; 1-dimensional, a line; 2-dimensional, like triangles, squares, circles, etc.; 3-dimensional, like cubes, spheres, and so on. Two-dimensional shapes are called polygons, while three-dimensional ones are called polyhedrons. As for those with 4-dimensions, they are called polytopes. The 2-dimensional polygons work as the basis from which all other higher-dimensional shapes can be formed. And just like prime numbers are considered the building blocks of all numbers, so are polygonal shapes with prime numbers for their sides' count, where prime polygons like the triangle, pentagon, heptagon, etc., can generate all other polygons as well.

Using dimensionality as a criterion is a powerful methodology as it extends to forms and shapes that are either regular, having some sort of symmetry (triangle, square, pentagon, cube, etc.), or non-regular, having no apparent symmetries in their forms. However, it has one drawback; it is unbounded, having as many values as we wish, 1-, 2-, 3-, 4- dimensions, up to infinite dimensions. So it is not like odd and even, or rational and irrational, where the space of categorization is limited to two or three. It is more like categorizing numbers by their values, 1, 2, 3, 4, …, all the way to infinity. Of course, in the case of numbers, this doesn't make any sense; however, for geometry, it does, probably because our perception is limited to three dimensions only, and therefore, extending the categories all the way to infinity is, by default, redundant and unnecessary.

Two Dimensional Polygons

Classifying 1- dimensional shapes is a straightforward matter, being mostly straight lines, finite or infinite in length. So we are not going to further investigate them here. However, by increasing the dimensionality to two, the number of classification' methods explodes. The most obvious method is through the number of sides. Each polygon is also unique by its interior angles. The interior angles of any triangle add up to 180°, always. The four angles of the square add up to 360°, 90° each. The interior angle of the pentagon is 108°, corresponding to an exterior angle of 72°, a number of utmost importance to geometry as we will discover. The five exterior angles of the pentagon sum up to 360°, the angular span of a complete circle, which is the case for the sum of the exterior angles of all polygons.

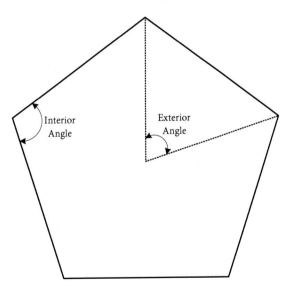

Figure 64: The definition of the exterior and interior angle of polygons. For the pentagon, the exterior angle is 72°, and the interior is 108°.

Below is a table listing the first ten regular polygons, along with the number of their sides and their angular configuration. The term *regular* refers to the those polygons with matching angles and equal sides. For example, the equilateral triangle of equal sides is the regular form of 3-sided polygons. Similarly, the square is the regular form of 4-sided polygons. Generally speaking, regular polygons are associated with equal sides and angles. Notice how the sum of the interior angles of all polygons is always a multiple of 180°.

Polygon	Number of Sides n	Interior Angle $[(n-2)/n] \times 180°$	Sum of Interior Angles $(n-2) \times 180°$	Exterior Angle $360°/n$
Triangle	3	60	180	120
Square	4	90	360	90
Pentagon	5	108	540	72
Hexagon	6	120	720	60
Heptagon	7	128.571428…	900	51.428571…
Octagon	8	135	1080	45
Nonagon	9	140	1260	40
Decagon	10	144	1440	36
Hendecagon	11	150	1620	32.72…
Dodecagon	12	162	1800	30

Table 18: List of some of the angular properties of regular polygons for sides from 3 to 12.

As obvious from the above table, the heptagon is a bit of an outsider whose interior angle is an infinite number instead of being finite like all other polygons. (Could the heptagon work as a wild card for geometry, just as it did with pure numbers?)

When the number of the sides n goes to infinity, the polygon will converge (or diverge) into a circle. The value of the exterior angles reduces to almost zero; however, their infinite sum remains bounded at 360°. The interior angles, on the other hand, converge to 180°, with the sum diverging into infinity.

Therefore, the triangle represents the first polygon, the first 2- dimensional configuration, while the circle stands for the last one, the final polygon, and between them, all other shapes fit in. They are the Alpha-Omega (A-Ω) of polygons (strangely, the triangle and the circle very much resemble the forms of these two Greek letters). Remember also that these two shapes, the circle and the triangle, are the hallmark of the *D*-space, with the *D*-circles and their triangulation of the numbers [1, 4, 7], [2, 5, 8], and [3, 6, 9].

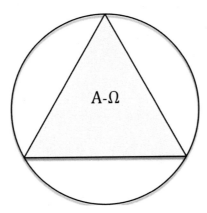

Figure 65: The triangle and the circle, symbolizing the beginning and the end, the finite and the infinite, the Alpha and the Omega.

Another way to distinguish between the different polygons is by how they can fill-up surfaces. For example, triangles, squares, and hexagons can fill-up the whole space by themselves only, with no gaps in between, as shown below.

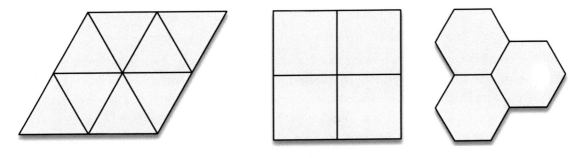

Figure 66: The triangle, the square, and the hexagon can fill up surfaces by themselves.

Starting with the pentagon, no polygon can fill up the whole surface by itself; it needs to be combined with others to perform the task, e.g., octagons with squares, etc. Nevertheless, the pentagon is unique as ten pentagons can form one complete circle, as shown below. And as we will discover next, the pentagon can form a closed 3- dimensional polyhedron, and so can the triangle and the square, while the hexagon cannot.

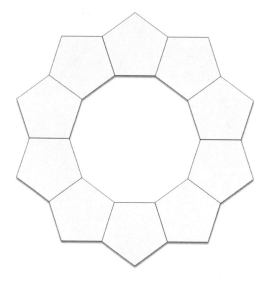

Figure 67: Ten polygons can form one complete circle.

Three Dimensional Polyhedra and the Platonic Solids

Adding another dimension to polygons transforms them into 3- dimensional polyhedra, where lines become surfaces, and surfaces transform into volumes. However, while regular polygons are all made of the same simple line added to itself *n* times, up to infinity, creating regular polyhedra from polygons is a bit trickier.

First of all, we do not have one simple shape that can create all polyhedra; we need to use different polygons, such as triangles, squares, pentagons, etc. Secondly, the process is not as simple as adding the same polygon to itself *n* times to create an *n*-faced polyhedron. For the polyhedron to close on itself and observe certain symmetries, the combination must follow certain rules and most often be made of different types of polygons, like in the soccer ball, for example, which is made of a combination of pentagons and hexagons.

There are few polyhedra with their faces made of the same polygon, referred to as *regular polyhedra*. There are only five (convex) regular polyhedra called *Platonic solids* (in contrast to the infinite possible regular polygons), shown below.

 The Platonic solids got their name after the famous Greek philosopher Plato (5th to 4th century BC), who associated these solids with the four basic elements: earth with the cube, air with the octahedron, water with the icosahedron, and fire with the tetrahedron. Later on, Aristotle, another Greek philosopher, completed the set by associating the ether with the dodecahedron.

95

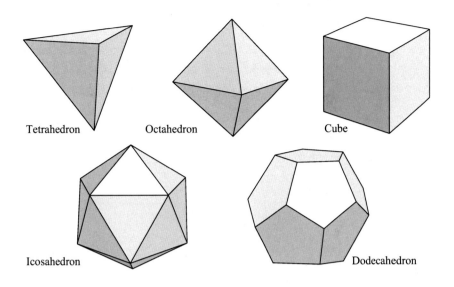

Figure 68: The five possible regular convex polyhedra that can exist, also known as the Platonic solids.

Each solid is made of one kind of polygons, whether triangles, squares, or pentagons. The one made of four regular triangles is called a tetrahedron. That which made of six squares is, of course, the cube. The octahedron is made of eight triangles; the dodecahedron is made of 12 pentagons, while the icosahedron is made of 20 triangles. Below is a table listing some of the properties of these solids.

Solid	Number of Faces	Number of Vertices	Number of Edges	D	Angles of Vertices	D
Tetrahedron	4	4	6	6	60	6
Cube	6	8	12	3	90	9
Octahedron	8	6	12	3	60, 90	6, 9
Dodecahedron	12	20	30	3	60	6
Icosahedron	20	12	30	3	60, 108	6, 9

Table 19: Properties of the five regular convex polyhedra, also known as the Platonic solids.

Notice the [3, 6, 9] digital root of the number of edges and angles of vertices. Also, note the correspondence between the number of faces and vertices for pairs of the solids, such as between the cube and the octahedron and between the dodecahedron and the icosahedron. These paired solids are called duals, as by connecting the centers of the faces of one shape, we generate the other.

For example, connecting the centers of the faces of a cube generates an octahedron and vice versa, as shown below. This is also the case for the dodecahedron and icosahedrons; they are each other's duals. This is because the icosahedron has 12 vertices, and the dodecahedron has 12 faces.

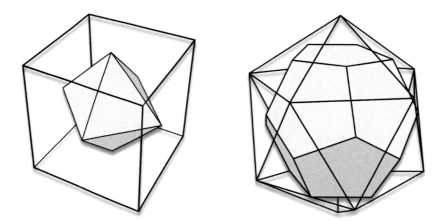

Figure 69: The dual properties of the Platonic solids, with the square being the dual of the octagon and the dodecahedron the dual of the icosahedron, and vice versa.

However, connecting the centers of the faces of a tetrahedron generates another tetrahedron, ad infinitum. Thus the tetrahedron is its own dual.

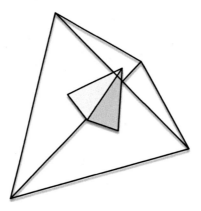

Figure 70: The tetrahedron: the solid that is its own dual.

The Polyhedra: An Expression of 720

Let us look at the sum of the angles of each polyhedron, just like we did for the polygons. To do so, we sum the angles at each vertex of the polyhedron and then multiply it with the number of vertices. For example, at each vertex of the tetrahedron, three triangles meet with the sum of the three angles as 60° + 60° + 60° = 180°. Thus, the total sum of the angles of the tetrahedron is 180° × 4 (vertices) = 720°. The table below lists the angular configuration for the Platonic solids along with their sum.

Solid	Sum of Vertex' Angles	Number of Vertices	Total Angles	Total Angles /720
Tetrahedron	180°	4	720°	1
Cube	270°	8	2160°	3
Octahedron	240°	6	1440°	2
Dodecahedron	324°	20	6480°	9
Icosahedron	300°	12	3600°	5

Table 20: The sum of all angles of the Platonic solids. It is always a multiple of 720.

Notice how all solids have their sum of angles divisible by 720°. This 720° dependency becomes much more apparent when we use the concept of *angle deficiency*, which is [360° − the sum of the angles at each vertex]. This value measures the defect of the angles to add up to 360°. When this value is multiplied with the number of vertices, we get 720° for all solids, as shown below for the Platonic solids, as well as for other not-so-regular polyhedrons.

Name of Solid	Number of Faces	Number of Vertices	Angle Deficiency	Total Angle Deficiency
Tetrahedron	4	4	180°	720°
Cube	6	8	90°	720°
Octahedron	8	6	120°	720°
Dodecahedron	12	20	36°	720°
Icosahedron	20	12	60°	720°
Truncated Tetrahedron	8	12	60°	720°
Truncated Cube	14	24	30°	720°
Truncated Octahedron	14	24	30°	720°
Truncated Dodecahedron	32	60	12°	720°
Truncated Icosahedron	32	60	12°	720°
Cuboctahedron	14	12	60°	720°
Rhombic Dodecahedron	12	14	77.88° (×6), 31.59° (×8)	720°

Table 21: The total angle of deficiency of all polygons is always 720°.

From the above table, the rhombic dodecahedron is of particular interest. It has a couple of angle deficiencies (77.88° and 31.59°) instead of one, as is the case for all other polyhedra. The reason becomes obvious when we look at its shape, shown below.

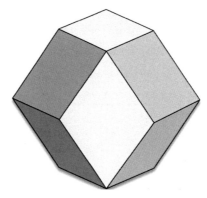

Figure 71: The rhombic dodecahedron, a unique shape in terms of its double angles of deficiency. It is also unique in that it can fill up space all by itself.

As shown, each rhombic face has two different angles: 70.53° and 109.47°, and they do not equally meet at each vertex. Four of the smaller angle meet at 6 vertices, while three of the larger ones meet at 8 vertices. Another unique property of the rhombic dodecahedron is that it can fill out space only by itself, similar to the cube, which is the only Platonic solid able to do so. (A combination of tetrahedrons and octahedrons can fill out space.) Its dual is the cuboctahedron, shown below, as obvious from their mirrored number of faces and vertices [14, 12]. The cuboctahedron is also a very special polyhedron where the distance from its center to each vertex is the same and is equal to its edges. This is why it is also referred to as *vector equilibrium*. However, the cuboctahedron cannot fill space by itself; we need to combine it with the octahedron to fill up gaps.

Figure 72: The cuboctahedron, dual to the rhombic dodecahedron. It has a unique property where the distance from its center to the vertices is equal to its sides.

Looking back on table (21), we notice the Platonic solids have the following multiples of 720: [1, 2, 3, 5, 9], so we are missing the [4, 6, 7, 8] multiples to complete the set to nine. We already found that the rhombic dodecahedron has a total angular sum equal to 4320 = 6×720, and therefore, it is the 6th multiple. The rhombic dodecahedron is the 7th, with the sum of angles 5040/720 = 7. But which polyhedra generate the 4th, and 8th multiples, corresponding to the sum of angles of 2880 and 5760? It turned out that the 8th multiple is a new geometry of an interesting configuration, shown below (hand-drawn by Robert Grant). It is based on the geometry of the cuboctahedron, but with some of the triangular faces turned into pentagons. Therefore, it is made of the main three polygons of the Platonic solids, triangles, squares, and pentagons (which correspond, numerically, to [3, 4, 5], being the dimensions of the first Pythagorean right triangle.) The remaining multiple need further research to be identified, that if it exists, of course.

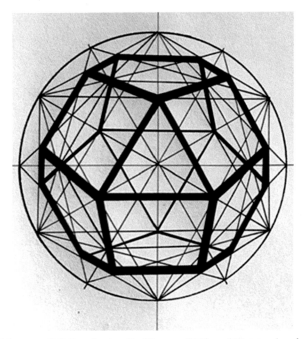

Figure 73: A Gyrobi-Cupola (Johnson solid) found using the Flower of Life and Metatron's cube to locate vectors. A new geometrical shape that embodies the square, triangle, and pentagon. It has a sum of angles equal to 5760, which is 8×720.

But why are all 3- dimensional polyhedra constrained to 720°? What is so special about this number? This we try to answer in the next chapter.

Number 72: The Trinary-Binary Number

To understand the meaning behind the number 720, we must return it to its roots, just like we do for real words. The root of 720 is number 72, which is derived from the anomaly numbers 2 and 3 by raising one number to the power of the other and then multiplying them together: $2^3 \times 3^2 = 72$. (We explained the anomaly status of numbers 2 and 3 when we investigated prime numbers.)

It can also be calculated from number 6 only, as $6^2 + 6^2 = 72$. In other words, it is the square of the hypotenuse of an isosceles right triangle whose sides measure six units each. Number 6 is the central number of our mathematical matrix, as we have discovered. No wonder that number 72, which defines 3- dimensional geometry, is related geometrically to it. And when we remember that number 72 defines the geometry of the pentagon, we get another interesting link between numbers 5 and 6.

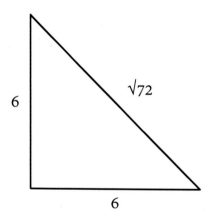

Figure 74: Number 72 can be defined through the geometry of an isosceles triangle whose sides equal to 6.

Numbers generated from the powers of 2 and 3 are called *binary* and *trinary numbers*, respectively. They start with 6, for powers of [1, 1], then 12 for powers [2, 3], 36 [2, 2], 72 [3, 2], and so on. Trinary/binary numbers are highly composite, having many factors compared to other numbers. They are the opposite of prime numbers, as we saw at the end of Part I. number 72 has 11 factors: [1, 2, 3, 4, 6, 8, 9, 12, 18, 24, 36]. Its reciprocal, however, has a decimal extension made of number 8 only: 1/72 = 0.013888... This is expected, as highly composite numbers do not exhibit a wave-like period in the decimal extensions of their reciprocals, which is another trait with which they oppose prime numbers.

Binary/trinary numbers are at the core of the Pythagorean Lambda, a triangular distribution of numbers believed to be very important to Pythagoras and his flowers. One side of the triangle is made out of the powers of 2, considered to be the female side, while the other is made out of the powers of 3, the male side, as shown below. By dint of this context, number 72 represents the perfect union of the male and female numeric aspects, similar to number 6.

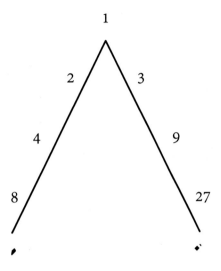

Figure 75: The Pythagorean Lambda is a distribution of male/female numbers, with the male being the powers of 3 (right side) and the female being the powers of 2 (left side).

The rhombic grid shown below is made from the multiples and divisions of number 72. This amazing grid series determines all *Time* numbers, the numbers that are the dimensional references of all space/time measurements using the imperial systems, and are fundamental to the scaling of each celestial body within our solar system.

These include the mile diameter measurements for the Sun, Moon, earth, and all planets, distances between the orbits of each, and the precession of the equinox. Each is derived as simple ratios of this magical number where 72 years is precisely 1 degree of the earth's precession cycle (72 x 360° = 25,920 years). Thus, number 72 doesn't only define 3- dimensional geometry. It also defines time, creating a sort of space-time unification.

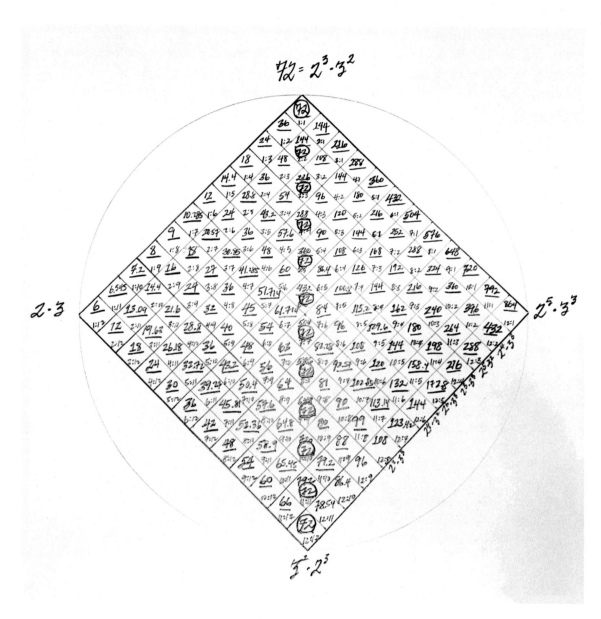

Figure 76: The grid (pattern) relationship between 2^n and 3^n. Number 72 is a special integer value due to its reciprocal relationship $(2^3) \times (3^2) = 72$. Several numbers on the grid appear in both esoteric and Kabalistic references to 72. Likewise, 7.2^2 = The Slope angle of the Great Pyramid at Giza (51.84°). (hand-drawn by Robert Grant). The resulting grid includes all the important numbers that correspond to measuring space and time.

103

All the above didn't seem to go unnoticed by the ancient scientists and engineers. For example, we find number 7.2 hidden in the square root of the slope angle of the Great Pyramid of Giza, the most magnificent edifice humans have ever built, $\sqrt{51.84} = 7.2°$. Also, the base length of the square in DaVinci's Vitruvian Man is 7.2 inches (while the top length of the square is 7.071 inches = $1/\sqrt{2}$), representing binary/trinary expansions of 72 at the point where the Vitruvian Man has his feet planted firmly on 'terra firma.' (More about the Vitruvian Man in Part V.)

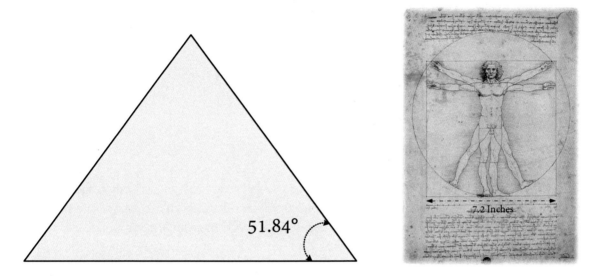

Figure 77: Left: The slope angle of the Great Pyramid of Giza is equal to the square of the number 7.2. Right: The Vitruvian Man of Leonardo Da Vinci has its feet resting on the side of a square whose length is 7.2 inches.

All of the above establishes the paramount position of number 72, as well as all of its fractals: 7.2, 72, 720, etc. Especially so, it is the number that represents the merger of the hexagon and pentagon as the 72° exterior angle of a pentagon matches the sum of angles of the hexagon, being 720°. This is, in addition, of course, to their right triangle relationship we just saw. And as we explained earlier, the two numbers that stand for angular and harmonic references, 360 and 432, are both multiples of 72, with the factors being 5 and 6, respectively.

And with 72 defining all 3- dimensional geometry, this number has the power to bring space and time, geometry and numbers, the pentagon and the hexagon, binary and trinary, as well as angular and frequency references, into a harmonious unity and wholeness. It is the most universal of numbers.

The Hexapentakis: The Geometry of Life

"Geometry is knowledge that appears to be produced by human beings,

yet whose meaning is totally independent of them."

-Rudolf Steiner

The Hexagon and the Pentagon: The Perfect Match

We have seen in the polyhedra section how some of those shapes are duals to each other, such as the cube and the octahedron; the dodecahedron and the icosahedron; and the cuboctahedron and the rhombic dodecahedron. This concept of pairing, however, doesn't have a 2- dimensional correspondence. Nevertheless, if any two polygons deserve to be considered dual or pair to each other, they would definitely be the pentagon and the hexagon. This is because the number of correspondences between these two shapes is absolutely unique from many perspectives, whether geometrical, numerical, mathematical, or even biological.

Geometrically speaking, a pentagon and a hexagon defined by the same circle will have an almost exact length for their sides (1.7% difference), as shown in figure 78, to the right. This is something unique among all polygons. Moreover, two identical circles, with the center of one lying at the perimeter of the other, form what is called the *Vesica Piscies*, which is one of the most profound and symbolic shapes in the human culture. These same two circles define an equal-sided pentagon and hexagon combination, as shown in figure 78, to the left.

When the hexagon and the pentagon are nested within two circles, if the inner circle's radius is set to the Euler number $e = 2.718...$, the outer circle's radius will almost equal to π (3.138..., about 0.1145% difference from real π), as shown in figure 79, to the right. And if we take the inner circle's radius to be equal to Φ, then the outer circle's radius will equal 1.868..., a value very close to the speed of light (*C*), in miles per second, around 186300 miles/sec.

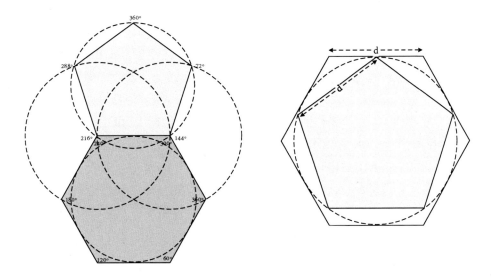

Figure 78: Left: The Vesica Piscies and a coupled hexagon-pentagon emerge from the same bi-circular configuration. Right: A hexagon and a pentagon defined by the same circle have almost identical sides.

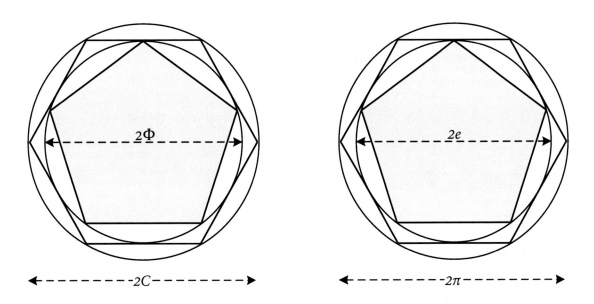

Figure 79: The hexagon and the pentagon are nested within two circles. Right: If the inner circle's radius is set to e that of the outer circle will equal π. Left: If we set the radius to Φ, then the outer circle's radius will equal to C = 1.868, the speed of light in miles/sec.

Getting e, π, and Φ from the geometry of 5 and 6 should not be surprising, as these fundamental constants are very connected to each other, numerically as well geometrically. For example, the π-root of e is $^{\pi}\sqrt{e}$ = 1.3748 ≈ 1.375, which is the golden angle. And we should not forget that 5 and 6 define π and Φ through their mutual ratio where $π/Φ^2$ = 6/5 = 1.2.

Geometrically, π and e define the relationship between a sphere and a cube inscribed inside it. We can prove it as follows. Let us take at a cube whose side is x. The diameter of this cube, which is the same as the diameter of the sphere that describes it, is equal to $2×r = \sqrt{3}×x$. This leads to $x = (2/\sqrt{3})×r$. The ratio of $2/\sqrt{3}$ is almost equal to the ratio of π to e (-0.0888% difference). Hence, $2/\sqrt{3}$ = 1.1547... ≈ π/e, and from this, the volumes of both the sphere and the cube will take the following values: $V_s = (4/3).π.r^3$ and $V_C = (4/3).(π/e).r^3$. Therefore, we can say that π governs circular and spherical geometries while e (along with π) governs linear, cubic ones.

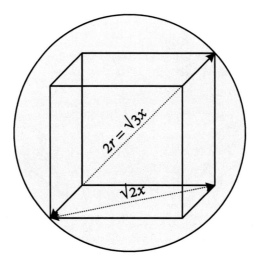

Figure 80: A sphere's volume is defined by the constant π. The volume of the cube it encloses, however, is defined by both π and e.

As we mentioned earlier, hexagons can form a honeycomb lattice that fills up a 2- dimensional surface completely. Pentagons can't fill up linear surfaces; however, ten pentagons create one complete circle, as shown below. Thus, the hexagon is linear in its essence, while the pentagon is circular, and together they can form an almost perfect sphere, like in the soccer ball. So we can say that the hexagon is related to e and the pentagon to π. No wonder when they are all connected via the relationship: $e×6+π×5 = 32$ (32.0176... to be precise). In fact, π and e satisfy another similar fundamental relationship: $e + 2π = 9$ (9.001467... to be precise).

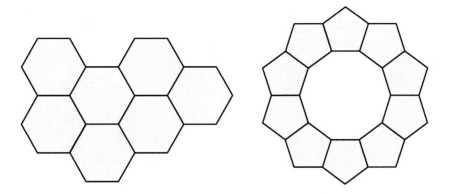

Figure 81: Left: Hexagonal lattice can fill up 2- dimensional surfaces with no gaps. Right: Ten pentagons create one complete circle.

Numerically speaking, and similar to number 72, numbers 5 and 6 have an interesting connection calculated from the first binary/trinary numbers of 2 and 3, through their product and sum: 2+3 = 5 and 2×3 = 6. Interestingly, in the *D*-space their product and sum reproduce numbers 2 and 3 back again: $D(6+5) = D(11) = 2$, and $D(6×5) = D(30) = 3$. We can also create a hexagonal configuration of 19 numbers, from 1 to 19, where the sum of the numbers in each direction equals to 38 = 2×19. Unlike magical squares, where there are many different versions with different numbers, sizes, and sums, this magic hexagon is a unique configuration that cannot be repeated using different sets of numbers. Notice how the number that occupies the central cell of this hexagonal configuration is number 5.

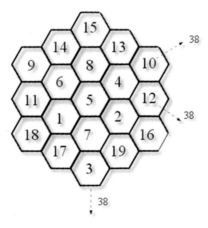

Figure 82: The Magic Hexagon: made of 19 numbers where the sum of the numbers in each direction equals 38. Notice the central position of number 5.

108

When it comes to nature and biology, the hexagon and the pentagon define the chemical compounds of the four nucleotides of the DNA, from which all life is created.

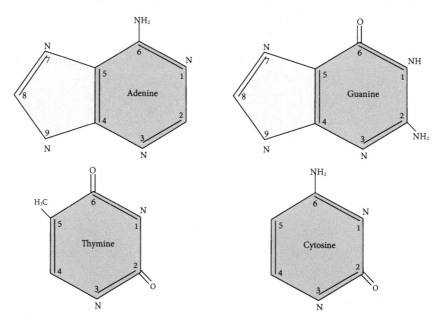

Figure 83: The four main nucleotides used in building the DNA helical strands of every living creatures are made of hexagonal and pentagonal rings of the elements Nitrogen N, Oxygen O, Carbon C, and Hydrogen H.

Interestingly, the three main units of the foot, the meter, and the cubit can be derived from the pentagon/ hexagon geometrical combination. As shown below, the foot unit emerges from the intersection of the pentagon and the hexagon, forming a right triangle.

Notice for the small circle, where the radius = 1 foot, 30° of arc equals 0.5236 foot. For the intermediate circle, where the radius = 1 cubit, a 30° of arc will equal 0.5236 cubits. For the larger circle, the radius is 1 meter, and 30° of arc equals 0.5236 meters. And a radius of 1.718 meters with 30° of arc equals 1.718 (e - 1) cubits. Likewise, the cubit and the meter are derived from fractal scaling and via dimensionless math constants, with 1 cubit = e -1 in foot measurement and $\pi/6$ in meter measurement, while the meter is $(6/\pi)$ ×(e - 1) in feet. It is the natural inward fractality at its next inward scaling.

Therefore, these basic units of measurements are not arbitrary and cannot be attributed to some historical consensus that rely on trivial attributions to man's foot size, or arm span, etc. Their measurements are the result of very precise and deliberate research based on a higher understanding of math, geometry, and cosmology. More importantly, they seem to have been contrived simultaneously.

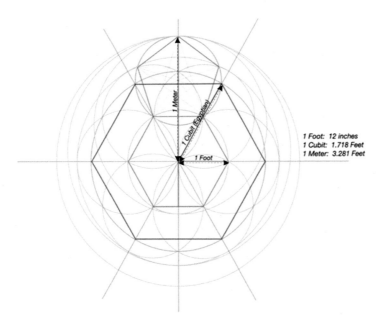

Figure 84: The imperial (foot), ancient (royal Egyptian cubit), and metric (decimal) systems of measurement appear to be inherent ratios of the Flower of Life's many circular intersections, requiring only minor adjustments as a viable combined system of measurement.

And speaking of measurements, one of Egypt's most iconic pyramids, the Bent Pyramid, is built based on a very unique and enigmatic blueprint that incorporates a precise pentagonal/hexagonal geometry. As shown below, the pyramid has two levels of masonry having different tilt angles, hence the name Bent Pyramid. The triangular face of the lower level has the exact dimensions of a 6th of a hexagon, while the upper level has the dimensions of a 5th of a pentagon.

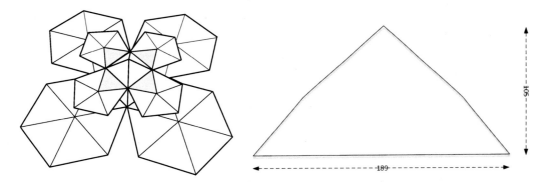

Figure 85: The Bent Pyramid of Egypt is built such that the dimensions of its lower level and those of its upper level reflect the hexagonal and pentagonal geometry precisely.

110

These very precise measurements cannot be attributed to mistakes by its engineers, as archeologists want us to believe. It is a well-intended and thoughtfully planned endeavor that resulted in one of the most amazing marvels of the ancient world. But, what could they be trying to tell us from this magnificent edifice? Probably all the above, and much more. Their understanding of these two shapes' importance must have been far ahead of ours for them to sacrifice all the time and effort to immortalize their measurements in stone. We are only scratching the surface.

Geometry in the *D*-Space

"EYPHKA! num = Δ + Δ + Δ!"

-Carl Friedrich Gauss

In this chapter, we investigate geometrical shapes from the numerical perspective of figurate numbers, as when dots or points are arranged in the desired geometrical form, their numbers will correspond to that specific shape. For example, the figurate numbers that correspond to a triangle, called triangular numbers, start from 1 (as all figurate numbers do). Then we add two dots to a total of three, then another 3 to a total of 6, and so on, as shown below.

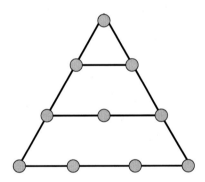

Figure 86: The first four triangular figurate [1, 3, 6, 10]

Figurate numbers can be 2- dimensional or higher. One interesting fact about 2- dimensional ones is they can work like prime numbers do, as any positive number can be written as the sum of a specific sequence of these numbers. This was proposed by the French mathematician Pierre de Fermat (1601-1665) to be later proven by his countryman Augustin-Louis Cauchy (1789-1857).

The German mathematician Carl Gauss was so excited when he realized that all numbers could be written as the sum of no more than three triangular numbers, that he wrote: *"EYPHKA! num = Δ + Δ + Δ!"* in his diary (*EYPHKA* is the Greek form of *eureika*).

In general, any positive number can be expressed as the sum of 3 or fewer triangular numbers, 4 or fewer square numbers, 5 or fewer pentagonal numbers, and so on. For example, 30 = 3+6+21 using triangular numbers or 30 = 1+4+9+16 using square numbers, etc.

Figurate numbers represent the perfect union between geometry and numbers. And when combined with digital root math, they transform into finite cycles that convey so much information about these shapes, much more than they ever would do in their original infinite format.

The figure below illustrates the *D*-circles formed from the digital roots of triangular, square, and pentagonal numbers. Notice how the square *D*-circle is made mainly of numbers [1, 4, 7] and number 9 while exhibiting a reflection symmetry. The pentagonal *D*-circle is made of all numbers from 1 to 9, with no reflection symmetry. The triangular circle is made of [3, 6, 9] and number 1 only, with reflection symmetry around the bottom 1.

Interestingly, these numeric observations apply to all polygonal numbers that belong to the same triplet group. So heptagonal numbers (7) will also exhibit the same numeric symmetries of the square, and octagonal numbers (8) follow the same numeric composition of the pentagonal *D*-circle, etc. These properties will be very useful later on when we investigate the correspondences these numbers have with some of the physical aspects of nature.

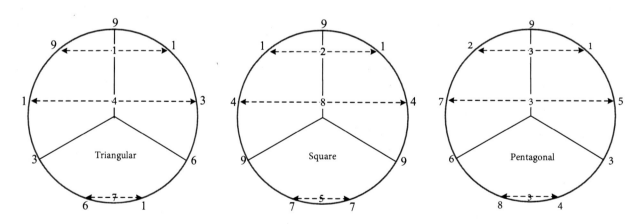

Figure 87: The numeric composition and symmetries of the *D*-circles corresponding to the triangular, square, and pentagonal numbers.

One interesting observation is that those polygons having the same digital root for the number of their sides will generate the same exact *D*-circles. For example, the 12-sided polygon has the same exact *D*-

circle of the triangle, as $D(12) = 3$. the same goes for 21-sided, 30-sided, etc. This is very interesting, as even when these shapes are very different from each other, having different symmetries and polarities, because their sides count for the same digital root, they still are treated as being the same by this operation.

We will investigate the 3- dimensional aspect of these numbers, specifically those corresponding to the Platonic solids, due to their unique geometrical and natural properties, which renders them an excellent archetype to examine their circular and symmetrical numerical properties within the D-space. As you may remember from chapter 1 of this part, the Platonic solids are made of five polyhedral: the tetrahedron, the octahedron, the cube, the dodecahedron, and the icosahedron. Below we conduct the digital root analysis of each one separately.

The Tetrahedron: The Simplest, the Strongest, and the Softest

Made of four equilateral triangles, the tetrahedron is the simplest shape among the Platonic solids. However, behind its simplicity lies a powerful configuration that resides at the core of most living and non-living objects.

On the molecular level, many atoms combine in tetrahedral shapes, including carbon-based molecules, which are the basic elements of living creatures. This is because the tetrahedron provides the optimum configuration for any four atoms to bond with the carbon atom, offering them stability with a minimum amount of bonding energy.

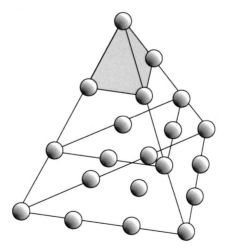

Figure 88: A graphical representation of the first four tetrahedral numbers: 1, 4, 10, and 20.

Diamond and quartz elements also have a tetrahedral shape. Both quartz and diamonds are the best natural heat conductors. (No wonder fire has been associated with this shape.) In contrast, water, the most malleable substance in nature, also has its molecules arranged in a tetrahedron shape. So this unique shape captures the essence of the hard and the malleable, heat and water.

In the D-space, the digital root of tetrahedral numbers reveals a 27-fold cycle, shown below. Notice the triple [3, 6, 9] axis of this circle, as if it tries to emphasize its 3- dimensionality.

Tetrahedral	[1, 4, 10, 20, 35, 56, 84, 120, 165, 220, 286, 364, 455, 560, 680, 816, 969, 1140, 1330, 1540, 1771, 2024, 2300, 2600, 2925, 3276]
D	[1, 4, 1, 2, 8, 2, 3, 3, 3, 4, 7, 4, 5, 2, 5, 6, 6, 6, 7, 1, 7, 8, 5, 8, 9, 9, 9]

Table 22: The digital roots of tetrahedron numbers repeat in the cycle of 27 digits.

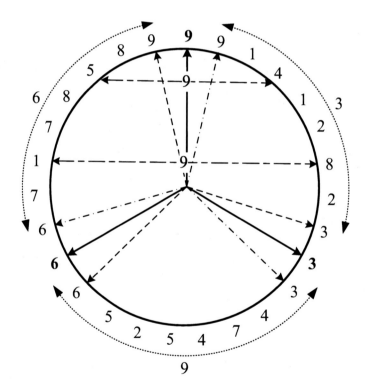

Figure 89: The 27-fold D-circle of the tetrahedral numbers, exhibiting a horizontal completion property (-sum of 9) as well as triple 3-6-9 axis.

115

The Cube: An Unearthly Shape

Cubic numbers can be generated graphically, as shown below. They go as [1, 8, 27, 64, 125, 216, 343, 512, 729, ...] Their digital roots generate a minimum of a 3-fold repetition cycle, made of the numbers: [1, 8, 9]. Not that interesting for a D-circle. However, these numbers are not as dull as they seem, as when adding them together or multiplying them by each other, they manage to reproduce each other back; 1+8 = 9, 1+9 = 1, 8+9 = 8, $D(8 \times 8) = D(64) = 1$, $D(9 \times 8) = D(72) = 9$.

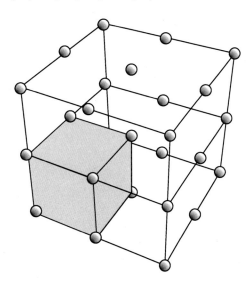

Figure 90: Graphical representation of the first three cubic numbers: 1, 8, and 27.

The cube represents the earth and the three-dimensional space that surrounds us. However, the cube is not a shape found abundantly in nature; it is too symmetric for it. It can be found in the symmetrical atomic structures of crystals but not in living things. Maybe this is why this shape was used extensively in sacred buildings of many ancient civilizations, symbolizing the power of man's order over the chaotic world of nature.

For example, Herodotus, the famous Greek historian, talked about an Egyptian shrine made of a single block of stone, measuring 40-cubits for all of its dimensions. The heavenly Jerusalem that John of Patmos saw in his dream had a cubical shape: "its length, breadth, and height were equal." And as we discover later on, the temple of Apollo at Delos was also a cube.

The Octahedron: The Phi Polyhedron

Similar to tetrahedral numbers, octahedral numbers generate a 27-fold repetition cycle in the *D*-space, as shown below, along with a triple [3, 6, 9] symmetry axis. Notice how their digital root starts with [1, 6, 1, 8], as in the golden ratio Φ = 1.6180…

Octahedral	1, 6, 19, 44, 85, 146, 231, 344, 489, 670, 891, 1156, 1469, 1834, 2255, 2736, 3281,3894, 4579, 5340, 6181, 7106, 8119, 9224, 10425, 11726, 13131
D	1, 6, 1, 8, 4, 2, 6, 2, 3, 4, 9, 4, 2, 7, 5, 9, 5, 6, 7, 3, 7, 5, 1, 8, 3, 8, 9.

Table 23: The digital roots of octahedral numbers also repeat in the cycle of 27 digits.

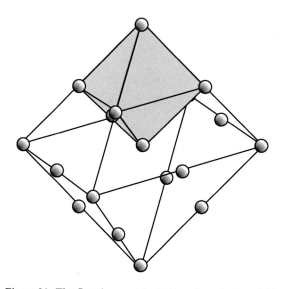

Figure 91: The first three octahedral numbers: 1, 6, and 19.

Octahedral shapes are found in the lattices of many of nature's crystals, where they develop into many different forms, called crystal habits. The most common habit for diamonds is octahedral in shape. So, in addition to these two Platonic solids (the tetrahedron and the octahedron) sharing the same 27-fold repetition cycle and the same triple 3-6-9 segmentations, they also share the same substance: diamond. Moreover, a lattice made of octahedrons and tetrahedrons can fill up space completely. And if we add the 27 digits of the digital root cycle of the tetrahedron, they add up to 135, which is the same number we get from adding up the 27 digits of the octahedron.

117

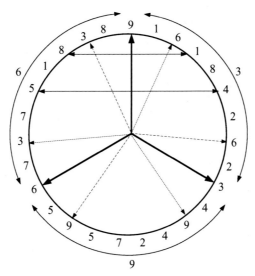

Figure 92: The 27-fold tetrahedral *D*-circle. Similar to the tetrahedral *D*-circle, this circle exhibit a horizontal 9-symmetry as well as a single 3-6-9 segmentation.

They are even more interconnected than that. If we chopped the sides of a tetrahedron in plains that are parallel to the tetrahedron facets, slowly, slowly, an octahedron would start to emerge. This operation is called *halving*. We can't do the same in reverse: chopping an octahedron to generate a tetrahedron. However, if we look at the same picture below, imagining that we are starting with an octahedron, we can generate the tetrahedron by extending the four facets of the octahedron that are tinted gray. When these surfaces intersect, they form a tetrahedron. The other four surfaces generate another inverted tetrahedron. In other words, when two inverted tetrahedrons are combined, they enclose an octahedron in between.

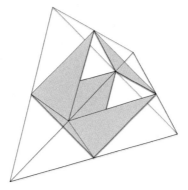

Figure 93: The halving process of the faces of a tetrahedron generates an octahedron. Extending the gray faces of the octahedron generates a tetrahedron.

118

So, these two shapes are inside-out reflections or mirror images of each other; one solid can generate the other and vice versa.

The Icosahedron: An Enigmatic Shape!

The icosahedron generates a 9-fold repetition cycle, as shown in the table below.

Icosahedral	1, 12, 48, 124, 255, 456, 742, 1128, 1629.
D	1, 3, 3, 7, 3, 6, 4, 3, 9

Table 24: The digital roots of icosahedral numbers repeat in the cycle of 9 digits.

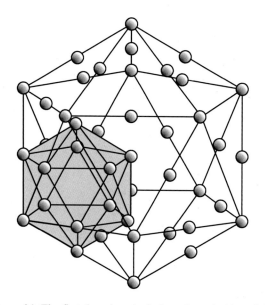

Figure 94: The first three icosahedral numbers: 1, 12, and 48.

Similar to the cube, we do not observe this shape in the natural world around us. However, powerful microscopes have revealed that the capsid (protein shell) of the AIDS virus has the shape of an icosahedron. As a matter of fact, most deadly viruses have capsids of either icosahedral or helical shapes. (Interestingly, Pythagoreans kept this specific shape a secret to themselves, considering it very dangerous if misused.)

119

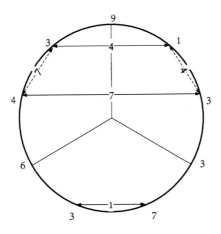

Figure 95: The 9-fold icosahedral *D*-circle, exhibiting [1, 4, 7] symmetries.

The Dodecahedron:

The first nine dodecahedron numbers are: [1, 20, 84, 220, 455, 816, 1330, 2024, 2925]. Surprisingly, the digital roots of the dodecahedron numbers generate exactly the same nine basic numbers, and in the exact ascending order: [1, 2, 3, 4, 5, 6, 7, 8, 9]. This hints at something fundamental about the dodecahedron, being the 3- dimensional archetype of the original number space. Recently, some scientists have suggested that the universe is finite and with a dodecahedral structure. And until recent times, the universe had always been considered being filled with ether. (Nowadays, it is believed filled with either vacuum or dark energy or matter.) Thus, it makes perfect sense when the dodecahedron was associated with the ether by the ancient philosophers.

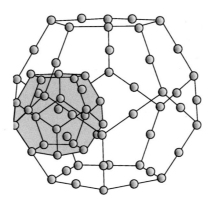

Figure 96: The first three dodecahedral numbers: 1, 20, and 84.

The icosahedron and the dodecahedron share the same repetition folds and are, interestingly, each other dual, as one can nest inside the other, as shown below.

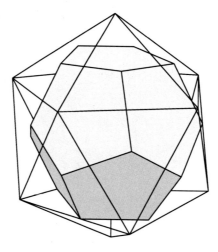

Figure 97: The dodecahedron is the dual of the icosahedron.

Plotting the Solids

In this section, we plot the D-numbers of the Platonic solids using Excel and see what information we can gain. At first, not much was observed from the individual plots of these shapes, as they produced asymmetric plots that do not seem to convey much information. Therefore we are not going to produce them here. However, looking at combinations of these solids reveals some interesting results. We consider different algebraic operations between the digital root numbers of these solids, such as addition, subtraction, and multiplication. The addition operation didn't produce anything interesting; however, the subtraction and product operations did.

Remember that the tetrahedron and the octahedron shared the same repetition fold of 27 digits, while the dodecahedron and the icosahedron shared a repetition fold of 9-digits. The cube repetition fold was 3-digits only. Therefore, it is logical to look for combinations of those polyhedra that shared the same repetition folds.

Starting with the tetrahedron and the octahedron, we were able to get interesting symmetrical plots only when we subtract them from each other, as shown below.

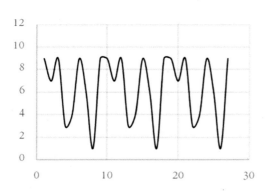 

Figure 98: Scatter (left) and radar (right) plots for the difference between the tetrahedral and octahedral *D*-numbers.

For the case of the dodecahedron and the icosahedron, it was the product operation that resulted in a symmetrical plot.

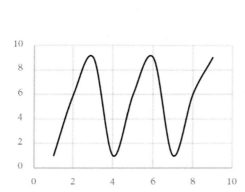 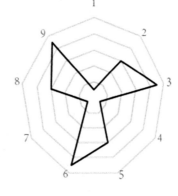

Figure 99: Scatter and radar plots for the product of the dodecahedral and icosahedral *D*-numbers.

The tetrahedron is the twin of the octahedron through the halving process, while the other two are each other duals through the matching of their faces and edges. This could explain the two different algebraic methods we had to use to generate the above symmetric plots.

The subtraction operation can be understood in the light that the halving process requires removing (subtracting) pieces from the tetrahedron to reveal the octahedron. For the case of the two dual shapes, the dodecahedron and the icosahedron, notice that the condition for being duals is that the faces of one equal the edges of the other and vice versa. Therefore, the products of these two properties for the same polyhedron are equal for both shapes. From the table below, we find that $12 \times 20 = 20 \times 12 = 240$. This is not the case for the tetrahedron and the octahedron.

Name of Solid	Number of Faces	Number of Vertices
Tetrahedron	4	4
Cube	6	8
Octahedron	8	6
Dodecahedron	12	20
Icosahedron	20	12

Table 25: Dual shapes share the product of their faces and vertices.

Even though the cube and the octahedron share the same product ($8 \times 6 = 6 \times 8 = 48$), their digital root numbers, for some reason, do not match their total: 27 for the octahedron versus 3 for the cube. There is a mystery here, as even though these two shapes are dual to each other, numerically, they are not that compatible. Symbolically, one of them stood for the earth while the other for air, two opposing elements that are hard to reconcile. The icosahedron and the dodecahedron, however, stood for water and the ether, and these two substances have always been considered highly connected. Could this be a reasonable explanation? Maybe, as numbers, symbology, and science are all intertwined on so many levels, we are only starting to comprehend.

Nevertheless, the above patterns and agreements definitely boost our confidence in the D-space in particular and in numbers in general. Every digit seems to be exactly where it should be, no more, no less. As they say, numbers do not lie, and in the D-space, they do not disagree either.

Semiprime Factorization
The Geometric Solutions

"The problem of distinguishing prime numbers from composite

numbers and of resolving the latter into their prime factors

is known to be one of the most important and useful in arithmetic."

-Carl Friedrich Gauss

The Sum-Product Approach to
Semiprime Factorization

Many public-key cryptography algorithms base their security on the assumption that the discrete logarithm problem over carefully chosen groups of numbers has no exact nor efficient mathematical solutions. One such important group emerges from the product of very large prime numbers, where finding the factors from the product is a very challenging and resource-consuming problem.

However, mathematical challenges often have geometrical aspects that better explain the problems, if not provide novel approaches to solving them. Consider Pythagoras's rule for right triangles. Mathematically, we know that the square of the hypotenuse is equal to the sum of the square of the other two sides. By involving geometry in the analysis, where the squared sides of the triangle are visualized as areas, the problem becomes better understood through illustration, which also generalizes it to all right triangles.

The search for the factors of semiprime numbers (or basically any number) is not only a mathematical necessity to satisfy the requirement of encryption or other related fields. On the contrary, reducing numbers to their basic units is similar to describing matter by the particles that make it. By studying this process, we may be unlocking much more than simple decryption techniques. In fact, we believe that factorization is probably a process used by all the elements of nature; it could be the most basic operation performed by electrons and atoms, as well as our own brains.

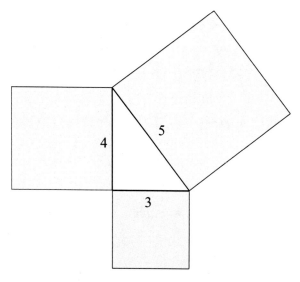

Figure 100: The geometrical aspect of Pythagoras's rule transforms the problem from an abstract mathematical formalism into a tangible two-dimensional Euclidean challenge.

In this section, we investigate the factorization problem by the geometrical analyses of the product and the sum of two prime factors. The problem of finding sum-product entanglements is an important one tackled by many mathematicians around the world. For our factorization investigation, we concentrate on finding the relationship between the perimeter and area of a rectangle, which has prime numbers for its dimensions, starting from a square of equal sides. We tackle the problem from two approaches. In the first one, we assume the product value is the area of a square (its dimensions are the square root of the semiprime value). We then try to find the dimensions of a rectangle that possesses the same area as the square. In the second approach, we require that both polygons have identical perimeter values instead of the same area values.

For each of these two cases, there are algebraic equations that describe the relationships between the parameters, depending on what is preserved during the transformation: the area or the perimeter. (There are more mathematical operations in this section than in any other one, but the math is still simple and straightforward to comprehend.)

Preserving the Area

In this case, the two areas of the square and rectangle are identical and equal to the product of the two prime numbers (x, y), which we label $A = x \times y$. The sides of the rectangle, x and y, are the prime factors we are searching for. For the A-square, we can find its sides (z) by taking the square root of area A. The perimeters of the two shapes are different, however, with the difference calculated as follows:

$$\Delta p = 2(x+y) - 4z = x + A/x - 2\sqrt{A}$$

Thus, the difference depends on A as well as on one of the sides of the rectangle, whether x or y. The relationship is also non-linear in regard to x.

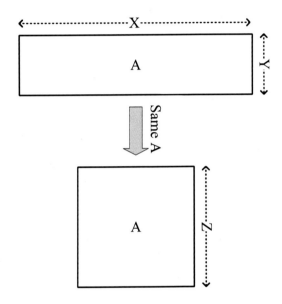

Figure 101: Squaring the rectangle while keeping the area constant.

Preserving the Perimeter

In this case, we hold the perimeter constant for both shapes while the two areas are drifting apart. The perimeter is calculated as $P = 2(x+y) = 4z$. The difference in the areas is calculated as follows:

$$\Delta A = A2 - A1 = z^2 - x.y = [(x + y)/2]^2 - x.y = [(x - y)/2]^2$$

As we are only given $A1$, we can rearrange the above equation to:

$$(x + y)^2/4 - (x - y)^2/4 = A1$$

126

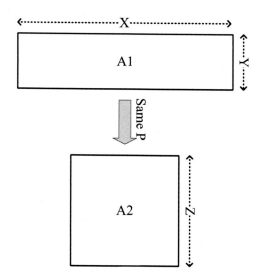

Figure 102: Squaring the rectangle while keeping the perimeter constant.

We now perform an in-depth analysis of the above two methods, which will reveal interesting geometrical correlations that enable the discovery of different powerful approaches to the semiprime factorization problem.

Lines of Constant Sum and Constant Product

We illustrate the geometrical correlations that may exist in the problem through examples. We choose our prime numbers to be 137 for x and 89 for y, and we use the multiplication table as our grid (the Q-grid). In the figure below, we plot two prime rectangles to illustrate the mirror symmetry of the geometry around the diagonal line of the multiplication plane ($A = x \times y = y \times x$). The prime rectangles have areas equal to A = 137×89. This value is represented by the tips of both rectangles.

The geometrical mean of the two prime numbers is calculated as the square root of their multiplication A. This is the length of the square z = $\sqrt{(113 \times 89)}$ = 110.421..., sharing the same area with the prime rectangles. The dimensions Rr and Rs are the diameters of the prime rectangle and the geometrical mean square (A-square), respectively. The P-line is the line that joins the two tips of the prime rectangles. This line is of significant importance to the problem. We add the average of the two prime numbers to the geometry in the form of a new bigger square, A_a, whose sides equal to (89+137)/2 = 113.

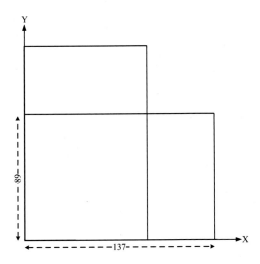

Figure 103: The geometrical representation for the case of A = 137×89 = 12193.

The image below illustrates adding the average square to the geometry of the problem, and the emergence of the *P*-line.

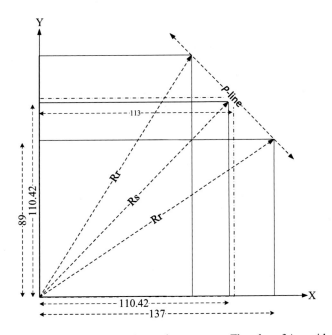

Figure 104: Imposing the average square Aa on the geometry. The edge of Aa resides on the P-line.

The right angle (corner) edge of the average-square lies exactly on the P-line, as by looking at the large right triangle whose sides are $x + y$, it is geometrically proven that the mid-point of the hypotenuse will project into the midpoints of the other two sides. This P-line is, therefore, the line where all points share the same value of $p = x + y = y_0$; it is the line of the constant sum.

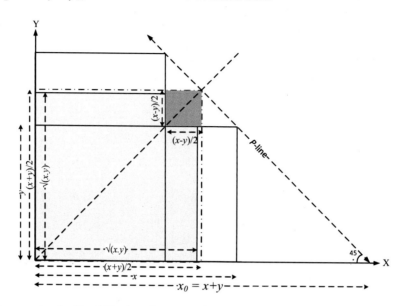

Figure 105: The P-line crossing the X and Y axis at the exact value of $x+y$. The dark shaded square represents the difference between the area-square and the average-square.

Theoretically speaking, knowing y_0 or x_0 enables us to figure out the two prime factors by substituting $y = y_0 - x$ into $A = x.y$ which leads to: $x^2 - p.x + A = 0$, where $p = x + y = y_0$. The discriminant of this equation has the value of $\Delta = p^2 - 4A$. The two prime factors can then be calculated from: $x, y = (p \pm \sqrt{\Delta})/2$. It is only when the correct (p) is used that the square root of the discriminant ($\sqrt{\Delta}$) will equal an exact whole number (with no decimal extension).

The area of the small strip between the two squares, the A-square, and the average square, is the term $[(x-y)/2]^2$, and it is equivalent to the small intersection square formed between the prime rectangles and the average square, as shown in the above figure (the dark-gray shaded square).

This square or strip's significance stems from being the exact area, when added to A, will transform it into the average square A_a. Notice that the average square always has an exact square root for its area, equal to $(x+y)/2$. The square root of the small gray square (or strip) is also exact and equal to $(x-y)/2$.

The above analysis allows us to factorize semiprime numbers as follows:

- Mark the point (on the product table), which corresponds to the semiprime A. For this illustration, we use $A = 12{,}193 = 137{\times}89$.

- Specify the hypothetical line, parallel to the real P-line, such that it touches the tip of the A-square.

- Advance the hypothetical P-line along the grid and measure the area of the newly formed square (the hypothetical average square).

- Calculate the area of the new square and see if it has an exact square root.

- If it does, we calculate the difference between the area of the new square and that of the A-square and take its square root.

- If the square root is exact with no decimal extension, this implies that our hypothetical P-line is now in the correct position on the product table and its intersection with the diagonal line is the point $((x+y)/2, (x-y)/2)$. From here, we can proceed to calculate the two prime factors of A.

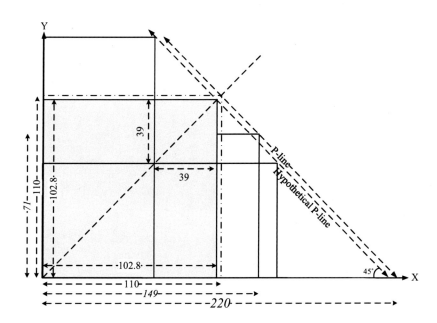

Figure 106: Finding the prime factors of the semi-prime number A = 149×71 = 10579 using the P-line method. By starting with a hypothetical line, we advance the line along the grid while calculating the area of the hypothetical average square. When the areas of the average square and that of its difference from the mean square have an exact square root, the P-line will be at the correct geometry corresponding to the two prime factors.

For the above case of 10579, the area of the correct average square is 12,100, which has an exact square root of 110. The difference between the areas of the two squares is 12,100 − 10,579 = 1,521, a number that also has an exact square root of 39. This means that the P-line is placed correctly on the grid, where the right angle (corner) edge of the two prime rectangles is situated.

In addition to the constant sum P-line, there is the line of the constant product, the A-line. This line has the equation of $y.x = A$, which is one form of a parabola curve, called a *rectangular parabola*, whose asymptotes are perpendicular to each other and where the principal axis, in this specific case, is described by the line $x = y$, as shown below.

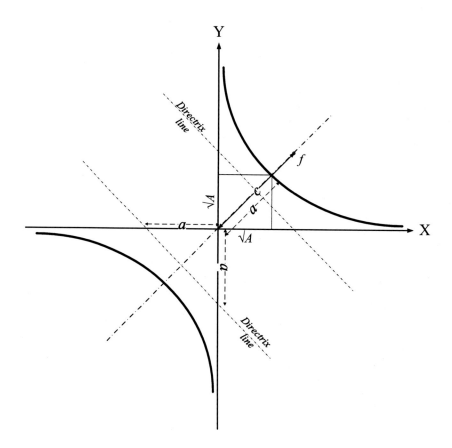

Figure 107: The constant-product line, the A-line, forms a rectangular parabola with its focus points residing along the diagonal line a distance c = 2√A from the origin.

The general equation of this parabola is $y.x = a^2/2$ with a being the semi-major axis. Therefore, a = √2A = Rs. The focus point of the parabola is located at point f, a distance c = 2√A from the origin. The eccentricity e of the parabola is calculated as $e = c/a = \sqrt{2}$. Consequently, the edges of the prime rectangles must fall at the intersection points (S) between the A-line and the P-line, as shown in the figure below. The A-line can be drawn exactly with its vertex (closest point to the focus f), coinciding with the right angle (corner) edge of the A-square.

When projected onto the Q-grid, the parabola can be used to factorize prime numbers. For example, a number like $A = 48427$ (613×79) has a digital root of 7; thus, only these combinations of digital roots for the prime factors are considered: (1, 7), (5, 5), and (2, 8). The last digit criterion restricts the two prime factors to possess the final digits of (1 and 7) and (3 and 9). Only these combinations, when multiplied, will result in the number 7 as the final digit for the semiprime 48,427. The qualifying numbers x_i are then tested such that when A/x results in an exact y_i, we know that we have found the correct prime factors. By calculating the A/x ratio, we are following the rectangular parabola line that satisfies the equation $y.x = A$. This method was very successful in factorizing public keys, especially when it was combined with the P-line method described above.

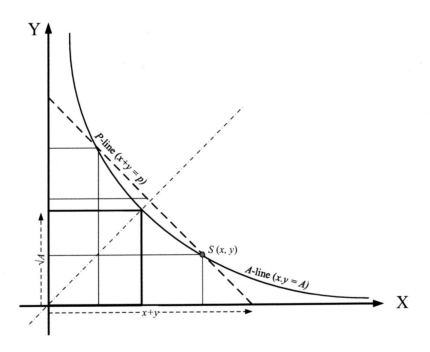

Figure 108: The intersection point of the A-line and the P-line corresponds to the right edge of the prime rectangle (point S). Calculating the ratio of A/x is geometrically equivalent to following the parabolic A-line, where $A = x.y$.

The Complimentary Roots

One of the equations we find from the above analysis, namely $(x+y)^2/4 - (x-y)^2/4 = A1$, is of particular interest. By rearranging its factors, we create an interesting relationship between the sum, difference, and product of the two prime numbers. By moving the difference term to the right of the equal sign and then substituting $A1$ with $[\sqrt{(p1.p2)}]^2$, the above equation transforms into the following form:

$$\left(\frac{x+y}{2}\right)^2 = \left(\frac{x-y}{2}\right)^2 + \sqrt{p1.p2}^2$$

This is simply a right triangle equation, with the two perpendicular sides being $(x-y)/2$ and $\sqrt{(p1.p2)}$. The hypotenuse is just $(x+y)/2$, as shown below. Notice how the side (or height) $\sqrt{(p1.p2)}$ is always an irrational number, while the other sides are always rational. Therefore, this equation not only defines the relationship between the sum, difference, and product, but also it is the equation that defines a unique geometrical relationship between rational and irrational numbers. Even though we derived the above relationship for the case of prime numbers and their multiplication, it is still not restricted to this case only; we can use whatever two numbers for x and y as long as they are not equal.

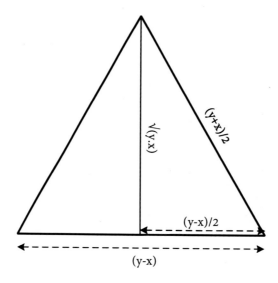

Figure 109: The relationships between the sum, difference, and product of two prime numbers (x, y) define a right triangle.

Interestingly, for the special case when one side (not the hypotenuse) is the square root of a quasi prime number (product of two primes, $p1$ and $p2$, both > 3), we discovered that while $C-B$ and $C+B$ (where $[C]$ is the hypotenuse, and $[A, B]$ are the other sides) are both irrational numbers, their product is always a

133

whole number, as shown below. We call these two numbers the *complementary roots*. These roots are different, but they are related to each other through their decimal extensions, being infinitely complementary, and always sum to one.

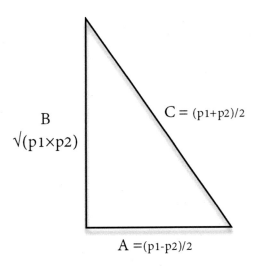

Figure 110: The right triangle of quasi prime numbers defined by the two primes numbers p1 and p2.

Consider the following quasi-prime number $30607 = 241 \times 127 = p1 \times p2$. It will yield a triangle with the following side lengths: $A = (241-127)/2 = 57$, $B = \sqrt{30607} = 174.9485638694985...$, and $C = (241+127)/2 = 184$, as shown below. The complementary root values are: $C-B = 9.051436130501488...$ and $C+B = 358.9485638694985...$

Notice how the decimal extension of $C+B$ is matching that of the square root of 30,607, and with the two decimal extensions of $C-B$ and $C+B$ summing to one, precisely: $0.051436130501488... + 0.9485638694985... = 1$.

Multiplying the two complementary root values together results in 3,249, a whole number that is indeed a perfect square value, as its square root is 57 (the precise value of side A).

One very interesting benefit of these complimentary roots is that they provide another method for factorizing semiprimes. We will not investigate this method further as it requires some technicalities that are not in the scope of this book. It is worth mentioning, though, that this method is unique as it only requires the semiprime number to be known beforehand and is successful in performing very fast factorizations.

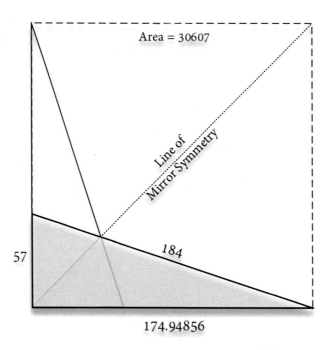

Figure 111: The right triangle for the quasi number of 30607. The three sides are entangled through the Pythagoras relationship as well as the complimentary roots.

In conclusion, transforming the abstract mathematical formalism of the factorization problem into a geometrical one, through the product-sum analysis, provided very powerful methods, proving once more that by combining math and geometry, we can make new discoveries, achieve better results, and gain a more profound understanding of the problem. Moreover, studying the geometry of one method may lead to another that is even more elegant, as is the case for the complimentary roots method mentioned above, along with their right triangle entanglement, which is going to be of fundamental importance to our understanding not only of mathematical problems but physical ones as well.

The Polygonal Intersection Method for Prime Factorization

This method of semiprime factorization is based completely on geometry. Its logic rests on finding the intersection points between polygonal shapes having areas equal to the public key.

The easiest way to demonstrate this method is through an example. We start with a public key of value A = 1891 whose two prime factors are 31 and 61. The first step is to draw a circle and a polygon, a square in this example, having equal areas of A. Next, another polygon of the same area A is drawn, which can be a triangle, a pentagon, a hexagon, etc., depending on the case. For this specific case, a triangle is chosen. The two polygons are centered around the center of the circle, as shown below.

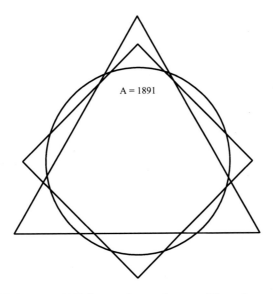

Figure 112: The value of the public key, A = 1891, is passed on as the area of three shapes, a circle, a square, and a triangle.

We rotate the triangle and the square by 30°, three times each, creating a 12-sided polygon, as shown below. Notice how the polygons intersect at specific points that do not touch the original circle. (This multiple rotation is only to make the intersection points clearer. We can definitely draw the circle from a single point of intersection.)

Figure 113: The square and the triangle rotated by an increment of 30°. They both intersect each other at specific points outside the original circle.

Next, a new circle is drawn such that it passes through the intersection points found in the previous step. The area of this new circle is measured to be equal to 2116, as shown below. The difference between the areas of the two circles is equal to $D = 2116 - 1891 = 225 = 15^2$.

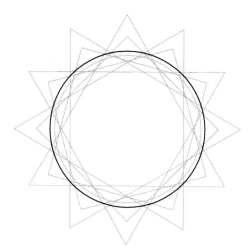

Figure 114: A circle is drawn such that it passes through the intersection points of the squares and triangles. The area of this new circle is measured to be 2116.

As it turned out, this value of 15 is exactly the difference between the mean (or average) of the two primes and the primes themselves, plus and minus. So, for a mean of $(61+31)/2 = 64$, the two primes are found from $61 = 46 + 15$ and $31 = 46 – 15$. This implies we need to find the average of the two primes beforehand to calculate their values from the difference D. However; this method has the advantage of enabling us to directly figure out the mean value as well, being nothing but the square root of the area of the new circle $\sqrt{2116} = 6$. These two numbers, 15 and 46, together with the square root of the public key (43.4856…) form a right triangle, with two sides having an integer value, while the third side ($\sqrt{A}$) is an irrational number by definition, as shown below. This is the same exact triangulation we discussed in the previous chapter.

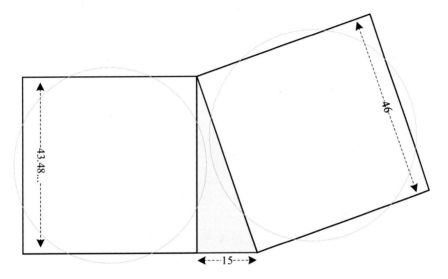

Figure 115: The mean, the difference, and the square root of the public key form a right angle triangle with two sides having integer numbers for their length, while the third is an irrational number.

Another example is for the semiprime $12193 = 89\times137$. For this case, we use a square and a pentagon instead of a triangle, as shown below. Using the pentagons and squares of area $A = 12193$, the two polygons intersect at points that delineate a new circle of area 12769, which is the square of 113. Their difference is equal to $D = 576 = 242$. From these two numbers, 113 and 24, we can find both prime factors: $89 = 113 – 24$ and $137 = 113 + 24$.

For a semiprime of $75349 = 151\times499$, the squares and triangles intersect at points that delineate a circle of area $A = 105625$, whose square root is $325 = \text{mean}(151, 499)$. For a public key of $323 = 19 \times17$, the mean circle is not found by the intersection points of a square and triangle. However, by using a pentagon, the intersection points land precisely at the correct value.

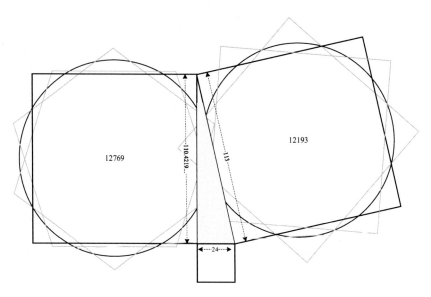

Figure 116: The case of A = 12193. By using squares and pentagons, the intersection points delineate a circle whose area is 12769 = 113², which is also the mean of the two primes 89 and 137. The difference between the areas of the two circles is 576 = 24², which is the square of the difference between the two primes and their mean: 89 = 113 – 24 and 137 = 113 + 24.

Sometimes, the difference between the semiprime circle and the mean circle is large, and the two polygons do not intersect at the correct position. This can be adjusted by multiplying the intersection circle by the factor of √2. This factor expands the intersection circle into the correct position of the mean circle, allowing us to calculate the values of the prime factors correctly.

The origin of the √2 factor is still not quite understood. Nevertheless, it helps to consider that when we take the ratio of the areas of the product circle $\pi.r_p^2$ and the average circle $\pi.r_a^2$, the value is equal to $(p1+p2)/2(p1 \times p2) = r_a^2/r_p^2$. Therefore, $r_a/r_p = (1/\sqrt{2}) \times \sqrt{[(p1+p2)/(p1 \times p2)]}$. The first prime number is 2, and for the special case where p_1 and $p_2 = 2$, the result is an exact 1 for $(p1+p2)/(p1 \times p2)$ and we are left with the √2 factor, which represents the maximum limit for the ratio between the two circles. All other ratios get smaller than √2 for any combination of numbers, whether they are equal or not. (Worth mentioning as well is the fact that the public key defines an area or an A-line, that is ruled by a rectangular parabola whose eccentricity is √2, as we explained earlier.)

The Flower of Life and Prime Factorization

The Flower of Life is a universal symbol deemed sacred by many civilizations around the world, including Egypt, Rome, India, and China. One engraving of this flower is found on the walls of the very ancient Osirian Temple in Egypt. The exact date of this temple is controversial, as some researchers date it back to many thousands of years, around 10,000 BC, much older than what orthodox Egyptologists are willing to accept.

The flower itself comes in many forms; all are based on overlapping circles, differing in their number, from 7 to 19 to 127, etc. The most famous form consists of 19 small circles encompassed by a large one, laid down in a hexagonal configuration, as shown below.

Figure 117: The Flower of Life, a mystical symbol made of 19 overlapping circles.

The Flower of Life is thought to embody the ferment of the original creation process, meant to represent the evolution of life from the rudimentary Seed of Life, made of the 7 central circles, shown below, into a fully developed human being. In fact, the seed of life has a very close resemblance to the shape of the first three mitotic divisions of the human embryo. It also represents the 7 days of creation, starting from the first circle for the first day, then two overlapping circles forming the Vesica Piscis for the second day, and so on until 7 circles are completely drawn.

Figure 118: The Seed of Life, found at the center of the Flower of Life, made of six circles encircling a seventh.

The flower is a very powerful blueprint, hiding within its proportions many important numbers and mathematical constants such as √2, √3, and the golden section Φ. Additionally, the central seed of life defines a [3, 4, 5] Pythagorean triangle, shown below. This right triangle is adopted by the designer of the second most important pyramid of the Giza Plateau, that of Khafre. If the base of this pyramid is scaled down to 6 units, its height reduces to 4 units. In other words, half the base, height, and slope form a [3, 4, 5] relationship.

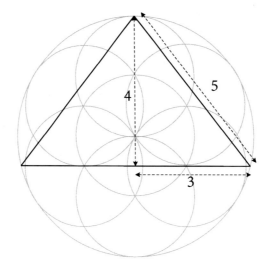

Figure 119: A [3, 4, 5] right triangle defined by The Seed of Life. The same triangle resides at the core of the Khafre Pyramid of Giza.

141

It is also a perfect embodiment of sacred geometrical forms, such as the Metatron cube, the vesica piscis, the hexapentakis, as well as the Tree of Life of the Kabala, among many others.

Figure 120: The Tree of Life embedded within the Flower of Life blueprint, delineating a pentagonal/hexagonal relationship.

In another rendition, the Flower of Life can be transformed into a torus-like shape (Torus of Life), shown below, with 24 intersecting circles and a 25th one at the center. What is so powerful about this shape is that it works like a measuring ruler where the intersection points of the 24 circles project on the horizontal line (the tangent to the central circle) units of equal distances, based on the units used in drawing the circles. The more circles we include in the torus, the more intersections we get and the finer the ruler's units become.

For example, if the central circle diameter is set to 4 units, which could be centimeter, inch, etc., the intersection points will project equal distances of 1 unit, 2 units, 3 units..., etc., as shown below. What is even more interesting is that for this specific value of the diameter (4 units), some intersections project an exact value for the Euler number and π, as shown in the same figure.

142

The number *e* vertical line goes through an intersection point that is 2.814 units from the center. The π-line goes through an intersection point that is 3.168 units from the center. Number 2.814 is the square root of earth's accepted diameter in miles, which is 7920 miles. The number 3168 is the length of the New Jerusalem, as proposed by John of Patmos, being 3168 cubits. What is more interesting is the fact that the circumference of the earth is circular and, through the Flower of Life, is related to π, while the New Jerusalem is said in the Book of Revelation to be cubic in form and is related to *e*. This agrees perfectly with what we suggested earlier in our sphere/cube analysis, with *e* defining cubic forms and π defining spherical ones.

This is another great example of the strong connection between science and ancient symbolism and how scientific knowledge used to be encoded within texts that do not at all pertain to such topics or themes.

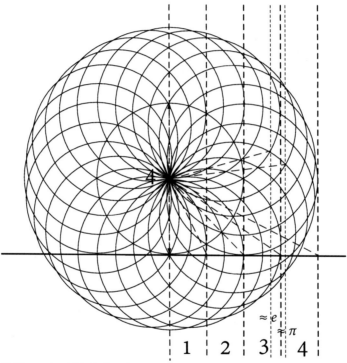

Figure 121: The Torus of Life: made of 24 circles plus a central one. It works as a ruler that can create equal distancing units as well as projecting values of mathematical constants.

The fact that these two fundamental constants emerge from a torus having circles' diameters equal to 4 units is not surprising, as 4 and π share a very powerful bond. For example, a circle with a diameter equal to 4 units will have an equal value for its area and perimeter, being 4π. This is a unique circle with an area-to-perimeter ratio of 1, a kind of boundary condition, where for values less than 4, the perimeter will exceed the area in value, and for values larger than 4, the area will exceed the perimeter. A square having a side of 4 units will have an equal area and perimeter as well, being 16 = 4×4. Thus number 4 defines circular (π) and linear (e) geometries having properties of perfect balance, no less than 1 and no more. Taking the ratio of the areas (or perimeters) of this special circle and square gives 16/4π = 1.2732... which is almost equal to the square root of the golden ratio where √Φ = 1.272...

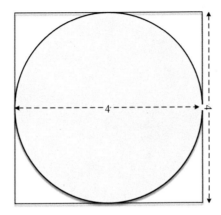

Figure 122: A circle and a square sharing the same dimension of 4 units will both have an area-to-parameter ratio equal to 1.

All the above suggests there is something important about number 4, especially as a measuring unit. Its ability to extract e and π from the Torus of Life as well as producing equal areas and parameters for the circle and the square is extraordinary. Add to it the fact that 4/π = √Φ, and we get something that definitely deserves to have us pause and ponder . And for the Pythagoreans, number 4 was the "fountain of nature" and its key. Being the first square number (2×2), a product of equal factors, 4 was the number of justice. Is it a coincidence when justice is about equality, and number 4 is where areas and perimeters become equal?

Squaring the circle was an obsession for ancient philosophers. Finding this exact value that can transform a square into a circle has always been an elusive endeavor due to the irrational value of the constant that governs circularity, the π number. But what if the real aim had always been to reconcile these two shapes into something meaningful, not exact, something similar to what we have discussed above with equal area and perimeters? And what if we squared the circles of the Flower of Life, using overlapping squares instead of circles? The emerging pattern forms overlapping Golden Rectangles across the flower's structure, as shown below. This is yet another form of the hexapentakis geometry because Golden Rectangles possess the same ratios derived from the square root of the number 5 (Pentagram and Pentagon), while the whole pattern is one large hexagon.

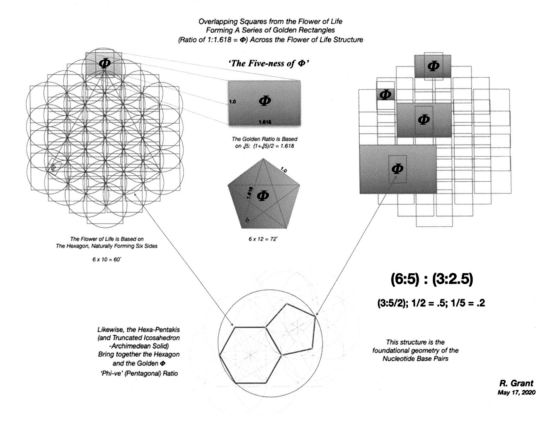

Figure 123: Replacing the circles of the hexagonal Flower of Life with squares creates a mesh of overlapping golden rectangles that embodies pentagons within their proportions.

Thus, we went on a full circle (or a square) from the Flower of Life to the Torus of Life to number 4 and the fundamental constants, to squaring the circle, to Φ and its golden rectangles, and ended with the hex-apentakis and back to the Flower of Life. What else can we learn from this amazing geometry?

The Torus of Factorization

The logic behind the polygonal intersection method for prime factorization we discussed in the previous chapter is not well understood yet. Luckily, the Torus of Life can shed more light on the geometric aspect of prime numbers, as it turned out the torus can also work as a geometrical factorization system.

From what we explained earlier, the intersections the circles make with each other set a scale, like a ruler, where the circles' diameters determine the units of that ruler. With only one starting value (any diameter value), we can derive the precise locations of whole integer Pythagorean right triangles and their respective side values by using the intersection points of the Torus of Life and its higher and lower "fractal" forms. This allows for prime factorization methodology using only one circle whose diameter is the square root value of the number whose prime factors we wish to identify.

Consider the case shown below. The number we wish to find prime factors for is $s = 253 = 11 \times 23$; therefore, the height of the right triangle (and the central circle diameter) is $\sqrt{253}$. The only right triangle that will possess two whole number values for both the base and the hypotenuse will have the following proportional dimensions: side A (base) = 6, and side C (hypotenuse) = 17 = $\sqrt{289}$. The two prime factors are derived as follows: $C - A = x = 11$; and $C + A = y = 23$. The circles' intersections signify at least two points that inform the precise line segment placement of the hypotenuse and, therefore, the other two sides as well.

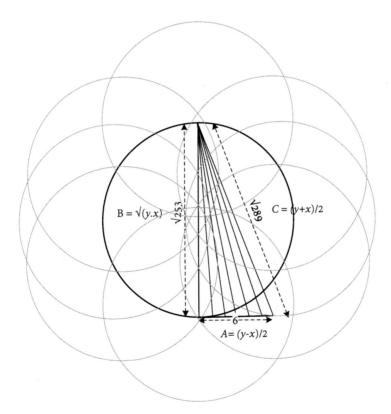

Figure 124: Using the intersection points of the Torus of Life to factorize a semiprime number. The diameter of the circles is equal to the square root of the semiprime. Only when the base, side A, and the hypotenuse, side B, are whole integers, can the prime factors be found from these two values.

All of this was obtained using only one equal diameter value (informing the length of side *B*) to locate the relevant circles' intersections and identify the other two sides of the right triangle and, by extension, identify its two prime factors (in a deterministic manner).

From similar examples as the above, we were able to deduce an equation that can factorize semiprime numbers perfectly, especially if the prime factors were close to each other. The equation goes as follows:

$$P_{x,y} = \sqrt{(s+\Delta)} \pm \sqrt{\Delta}, \text{ where } \Delta = round[(roundup(\sqrt{s})-\sqrt{s})/(\sqrt{(s+1)}-\sqrt{s}), 0.5] = A^2$$

Where the minus sign is used for finding one prime factor (*x*) and the plus sign is used for finding the other factor (*y*). The *round* function is used to round to the nearest half-integer. For the case where the prime numbers are far apart, the Δ function should be modified such that its value is not necessarily the square of the base of the triangle (*A*) but a multiple of it. The correct multiplication factor can be determined by through trial and error.

Being able to perform factorization from the semiprime alone via a simple equation, even if it is not 100% accurate for all cases, is by itself a major breakthrough in the field of prime numbers, as well as a further step in understanding the relationship these numbers have with geometry, which, by now, has become clear is based on the right triangles that emerge from the circular intersections of the Flower/Torus of Life. A triangle so simple and basic in its design, yet it embodies some of the most fundamental principles nature uses in its dynamic.

Simplicity is always the key to understanding.

The Right Triangle: The Marriage of Numbers

We all agree by now that the right triangle is a very fundamental shape, especially when it comes to semi-primes and their factors. That being said, the right factorization relationship is not reserved for prime numbers only; any number can be expressed through this relationship, as shown in the table below for numbers from 1 to 7. Of course, most numbers have many factors besides the ones shown below; however, only those that are defined by the Pythagorean rule exhibit such strong geo-numeric entanglement between them.

Number n	Factors n = x×y	Right-Triangle Sides [A, B, C]
1	[1, 1]	[1, 1, √2]
2	[√3+1, √3-1]	[√2, 1, √3]
3	[2+1, 2-1]	[√3, 1, 2]
4	[2, 2]	[2, 2, √8]
5	[√25, √25]	[√5, 2, 3]
6	[√10+2, √10-2]	[√6, 2, √10]
7	[√8+1, √8-1]	[√7, 1, √8]

Table 26: The first seven numbers and their corresponding right triangular representations.

One such unique triangle is shown below. Besides having its sides equal to the square roots of the [3, 6, 9] triplet group, taking the sum and product of its two irrational complimentary roots results in number 6, for both. As we showed earlier, the complimentary roots of this specific triangle can be calculated as √9+√3 = 4.73205... and √9-√3 = 1.26795... Interestingly, 4.73205... + 1.26795... = 4.73205... × 1.26795... = 6. Number 6 is considered a perfect number as the sum and product of its divisors are the same and equal to the same 6: 1+2+3 = 6 and 1×2×3 = 6, which renders this triangle a double-perfect one.

Actually, our research suggests that using these triangles to factorize prime numbers is something each one of us is constantly doing in our unconscious minds, where the complex factorization mathematics needed for many vital operations, such as vision, are performed

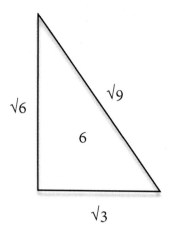

Figure 125: The perfect right triangle of sides equal to the square roots of the triplet group of [3, 6, 9] has the unique property where its two complimentary roots add and multiply to the same result of 6.

.

In the vision process, for example, the eye will increase its lens diameter by relaxing the ciliary body muscles in its anterior chamber. The lens diameter adjustment will also be accompanied by the unconscious modulation from a convex lens configuration (bulging lens at the center for near vision) towards a concave lens one (at higher diameters for distance vision).

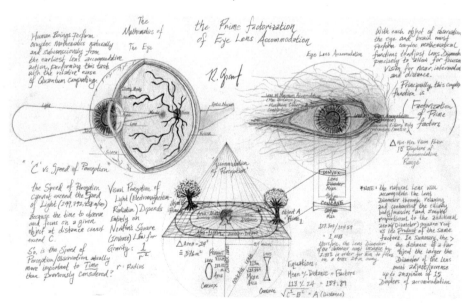

Figure 126: The prime-factorization geometric method used by the brain to achieve optical focus in the eye (hand-drawn by Robert Grant.)

149

Let us consider an observer who sets his gaze on a tree a distance r from where he is positioned, as shown below for the solid circle. In order to adjust natural lens diameter precisely to look on a further-out object, a distance R, the eye and unconscious mind must solve the R diameter of the mean (area) by finding the two prime factors (x and y) whose $\sqrt{(x \times y)} = r$, such that they make a perfect right triangle whose side is R (or nearest exact value). This, of course, takes some time complexity, which is approximately the same time it takes our eyes to refocus. So, for a distance $r = 43.4856...$, the two prime factors that form a perfect right triangle with a hypotenuse of 46 are 31 and 61.

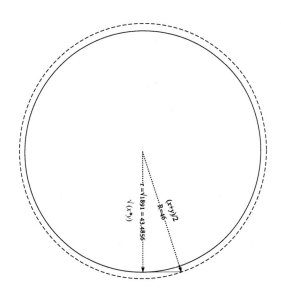

Figure 127: The area of the solid circle (r) represents the eye's current range of focus. The dotted circle (R) represents the new range of focus. Both create a right triangle based on prime number x, y such that r = √x.y, and R = (x+y)/2.

When two prime numbers (x and y) are multiplied, are they likewise added in some other dimension? What about their difference? The discovery that these are simple separations distributed across Pythagorean triangular dimensions opens the doors for more interesting discoveries that pertain to all fields of physics, from quantum mechanics to mathematics and even biology. In fact, the relationship between the electron, the proton, and the neutron could be the physical manifestation inherited from right triangle principles, as well as space-time, gravity and electromagnetism, and all other trinary relationships that we most often find in the laws that govern nature.

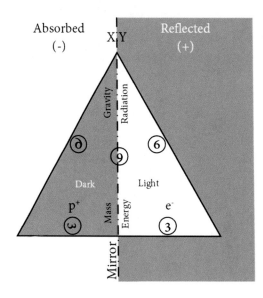

Figure 128: The triangular relationships between the various aspects of the physical world.

Besides their fundamental numerical relationships, right triangles can create some of the most important shapes in nature: spirals. It is already known that placing right triangles next to each other such that one triangle's hypotonus works as the other's perpendicular side will produce a perfect spiral for the special case when the outer sides of all triangles are set to 1 unit in length, as shown below. This spiral is called the spiral of Theodorus, in reference to Theodorus of Cyrene (340 – 250 BC).

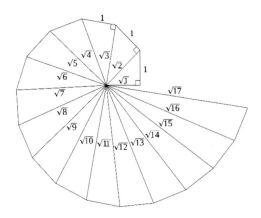

Figure 129: The spiral of Theodorus.

Now, if we constrained the height of the right triangles to be equal to 1, 2, 3… up to infinity, keeping the base at 1 unit, a new type of spirals emerges, shown below. This new spiral was named the spiral of Regulus by Robert Grant.

Thus, right triangles could be one basic method through which nature creates these wonderful shapes, and on all levels, from the largest macrocosm, such as clouds and galaxies, down to the smallest microscopic level of the DNA strand.

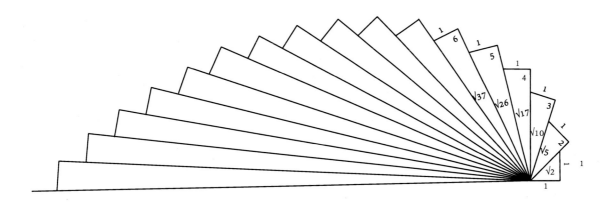

Figure 130: The Spiral of Regulus. All the bases are set to 1 unit. The hypotenuses range from 1, 2, 3, up to infinity. The heights are square roots of 2, 5, 10, etc.

One of our prominent colleagues, the Shakespeare scholar, Alan Green (www.ToBeOrNotToBe.org), discovered, encoded within the title page of the 1609 publication of Shakespeare's sonnets, a previously unknown connection between the primary Pythagorean triangle (sides [3:4:5]) and the three most significant mathematical constants π, Φ and e. Startling in its simplicity and elegance, he found that $[\pi/(\Phi \times e)] \times [(3+4)/5] = [\pi/(\Phi \times e)] \times [(\sin(\theta) + \cos(\theta)] = 0.99999 \approx 1$, where θ is the angle between the [3:5] sides. The full significance of this equation will be much more appreciated when we discuss the 12-base system and the three new hidden numbers.

Thus, the simple yet potent shape of the triangle is the ultimate encoder, a gateway through which our consciousness can stretch toward higher understandings. The perfect spirals it makes are the universal carriers that transcend all boundaries and limitations and through which all information can be coded and transported, right from the smallest DNA and up to the farthest galaxies.

The Fractal Root

"Maybe you are searching among the branches, for what only appears in the roots."

-Al-Rumi

The Square Fractal Root

Finding the roots of numbers, such as their square roots, is one of the most fundamental mathematical operations that have been implemented for millennia. It is an indispensable tool for solving many important mathematical problems, either algebraic or geometric, such as polynomial equations, standard deviations, the Pythagoras rule for right triangles, etc. It is also embedded within the formulas of some important physical constants, such as Planck's length: $p_l = \sqrt{(\hbar G/C^3)}$ (where $\hbar$ is the Planck constant, G is the gravitational constant, and C is the speed of light).

One very familiar root is the square root of number 2, which solves for the hypotenuse of a right triangle whose perpendicular sides measure to unity. Using Pythagoras rule, we can calculate the hypotenuse as: $\sqrt{(1^2 + 1^2)} = \sqrt{2} = 1.41421356...$ The irrational nature of this number is believed to have been confirmed by the Greek mathematician Hippasus of Metapontum (530 - 450 BC). (This discovery sent shockwaves throughout the Pythagoreans' society who would not accept the existence of such imperfect numbers (e.g., irrational, transcendental, etc.) to the extent they are rumored to have thrown Hippasus into the sea as a punishment for his heretical discovery.)

It is a consensus among mathematicians that any number's principal roots must be numerically identical in both: their numbers sequencing and relative magnitude. They can be negative or positive, but they must be identical in all other aspects. This restricted the root of any number to one value only, depending on the order of the root. For example, the square root of number 4 can be number 2 only, plus or minus. Similarly, the cube root of number 64 is ±4 only.

But could there be another way to square- or cube-root numbers? In other words, is it possible to find identical roots for any number, other than its formal roots, such that when multiplied together, they will generate the same number back? At first inspection, this question may seem ridiculous. Nevertheless, there is the possibility of a reasonable answer, given that we relax the constraints imposed on what we define as *identical* numbers.

As we explained above, the formal definition for two identical numbers is they have the same exact numerical sequencing and relative magnitude: their decimal points. For example, 3.14 is not considered identical to 0.314 nor to 31.4, etc. But as you may remember from Part I, the wave nature of constants showed us that we are allowed to take a bit of a relaxed approach to this strict definition by removing the decimal point constraint. In other words, we consider numbers like 3.14, 0.314, and 31.4 to be *fractally* identical. For example, a number like $\pi = 3.14...$ can also be written as 0.314×10 or $31.4/10$, etc. In other words, what really identifies this number is its unique numerical structure: 3, followed by 1, then 4, and so on; it is this specific sequencing that sets this number apart from all other numbers, not its decimal point. Working with this *fractally-identical* scheme of numbers expands our understanding of some mathematical operations while enabling the discovery of new ones. And one such operation is the *root*. We start investigating its fractal-based nature by examining the simplest case of the square root.

The square root reduces any number (y) to a pair of identical smaller numbers (x) such that $x \times x = x^2 = y$. The result can be an integer or a non-integer (a float), depending on the original number. We can also reverse the notation to write: $^2\sqrt{y} = x$. The number x is called the *principal square root* of y, which is called the *radicand*. The radical symbol ($\sqrt{}$) indicates taking the root of the radicand y, and the superscript "2" indicates taking the square root, not the cube or fourth roots, etc. The radical symbol may have its origin from the Arabic letter 'ج' phonetically equivalent to *G* or *J*, which is the first letter in the word *'jathr'*, meaning *root*, which was used by Arab mathematicians as a symbol for the root operation.

One peculiar square root is for number 10: $\sqrt{10} = 3.162277...$ This is because the inverse of this number is a fractal of the same exact number $1/\sqrt{10} = 0.3162277...$ We can approach the above from a different perspective involving the number 1 rather than 10, where $3.162277... \times 0.3162277... = 1$.

The formal solution to $\sqrt{1}$ is the two, numerically identical, principal roots ±1. However, we just found two fractally identical numbers, differing only by a decimal point, such that when multiplied together, they produce 1 also. To find this unique root of number 1, we multiply it by 10, then we find its principal square root p. The number 1 will then be equal to $p \times p/10$. This unique root of 1 is capable of reproducing the same kind of rooting for any other number. For example, taking number 5 as the radicand, we find $\sqrt{(10 \times 1)} \times \sqrt{5} = \sqrt{50} = 7.07106...$ and $7.07106... \times (7.07106.../10) = 5$.

We call this new root the *Square Fractal Root*, and we use the radical fractal sign $^{fr}\sqrt{y}$ to indicate its implementation. We also call the value √(10×y) the *Greater Fractal Root* (GFR) of *y* and the other root (GFR/10) the *Lesser Fractal Root* (LFR). Both roots can be either positive or negative similar to the principal square root. In fact, these two square roots, the principal and the fractal, are the only square roots a number can have, depending on its order of magnitude. So, for number 5 again, √5 = 2.2360... But √0.5 = 0.707106... = LFR(5), √50 = 7.07106... = GFR(5), √500 = 22.360..., etc.

Thus, the fractal root is a mathematical operation that introduces a different perspective to what we mean by the square root, generating two roots instead of one, with both being identical but for the position of their decimal points. Below is a list of the GFR of some constants of interest.

Number	Value	GFR
Pi π	3.140...	5.6049...
Euler *e*	2.718...	5.2137...
Golden section Φ	1.618...	4.0224...
Number 2	2	4.4721...
Fine structure α	137	37.013...

Table 27: The square fractal roots of some of the most important constants of nature

The square fractal root can provide different methods to derive some fundamental physical and mathematical constants. For example, adding 0.3 to the LFR of 1 results in a number almost identical to Planck's length $(1.616229 \times 10^{-35}$ m) minus 1: $(^{fr}\sqrt{1}+3)/10 = 0.6162277...$ This Planck's length is defined as the unit measurement that light will travel in one unit of Planck's time $(5.39 \times 10^{-44}$ s). It is also the scale at which quantum gravitational effects are believed to become noticeable. Additionally, the fractal roots enable a geometrical representation of this fundamental constant that is based on a circle having a diameter of 6-units and defined by the two values of $^{fr}\sqrt{1}$ - 3 and $^{fr}\sqrt{1}+3$, as illustrated in the figure below. (And again, we find number 6 playing a major role in defining constants, as it did in the wave theory of constants.)

Another usage is to calculate π from the LFR(1) as follows: $\pi = (^{fr}\sqrt{1} + 1)^{1/0.24}$ The offset from the real value of π is less than 0.015%. Additionally, we can calculate *e* from the fractal roots of 0.718, 2.68... and 0.268..., as follows: $e - 2 = 2.68... \times 0.268...$ We can also calculate the principal square root of 3 from the LFR(1) as follows: $(LFR(1) + 1)^2 = 1.732... \approx \sqrt{3}$.

One very important relationship in math is the Euler identity, defined as $1 + e^{i\pi} = 0$, where *i* is the imaginary number, the square root of -1. Interestingly, if we replace *i* with the lesser fractal root of 1, and take the inverse of the *e*-fraction, $1 + 1/(e^{0.316...\times\pi})$, the answer to the above equation will equal to 1.370293, almost identical to the fine structure constant. This is one form of Euler identity that is no less important than the familiar one, if not more important.

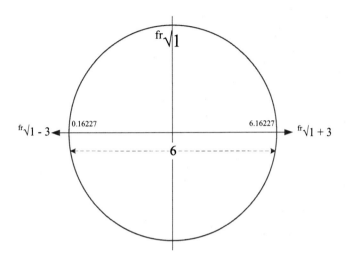

Figure 131: Planck's length, calculated from the fractal root of 1, is defined by a circle of 6-units in diameter.

Geometrically speaking, the square fractal root maps one shape to another, where both shapes have identical areas but with different dimensions. In other words, for some area A, using the principal square root, we assume the area to be that of a square with the sides having equal values (a ratio of 1). By using the fractal root, however, we assume the shape to be of a rectangle with its sides fractally identical, having a ratio of 10 instead of 1.

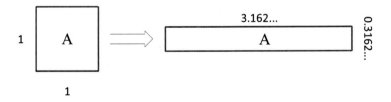

Figure 132: Using the fractal root to transform a square of area A into a rectangle having the same area but with the ratio of the sides equal to 10.

For the case of a circle, the area is calculated as $A = \pi \times r^2$. Using the principal square root, the radius is calculated from the area as $r = \pm\sqrt{(A/\pi)}$. The fractal root, on the other hand, generates two fractal values of *r*. In other words, we are assuming the area *A* to belong to an ellipse with the major and minor axis (*a* and *b*) being the GFR and LFR of *A*, respectively: $a, b = {}^{fr}\sqrt{(A/\pi)}$.

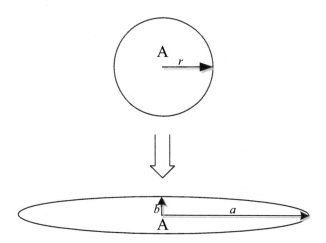

Figure 133: The fractal root transforms a circle of area A into an ellipse of the same area with a/b = 10.

Even though the areas stay unchanged for both shapes, before and after mapping, the perimeters do change. For example, in the case of the square with unity dimensions, its perimeter is $4\times1 = 4$. However, the fractally mapped rectangle has dimensions of (3.162... and 0.3162...), producing a perimeter of: $(3.162... + 0.3162...) \times 2 = 6.9570... > 4$. With the dimensions increasing in value, the perimeters of the square-shaped polygon get bigger in comparison to the rectangular fractal one, as shown in the figure below.

The perimeters' ratio will equal 1 for the single case of a square's dimension of the magical value of 3.025 units, calculated as follows. For a square of sides *x* and a perimeter P_p and fractally mapped rectangle of sides $\sqrt{(x/10)}$ and $\sqrt{10x}$, and a perimeter P_f, we find $P_p = 4\times x$ and $P_f = [\sqrt{(x/10)} + \sqrt{10x}]\times2$. Now, for $P_p/P_f = 1 = 4\times x/([\sqrt{(x/10)} + \sqrt{10x}]\times2)$, therefore, $2\times x = \sqrt{(x/10)} + \sqrt{10x}$, and $2\sqrt{x} = \sqrt{(1/10)} + \sqrt{10}$, which leads to a value of $x = 3.025$.

For the case of a circle and an ellipse, the magical value is $r = 5.05$ units. Interestingly, for dimensions equal to the fundamental constants π and *e*, the ratios are 1.0191 and 0.948, respectively, both very close to unity. (Also, notice how the digital roots of 3.025 and 5.05 are equal to 1.)

There could be some properties for the fractally mapped shapes that make them special to nature in one form or another, similar to those based on the golden ratio Φ, for example. As we have learned in Part I, the golden ratio division ensures that the ratio of the whole to the bigger part is equal to that of the bigger

157

part to the smaller one, as shown below. A rectangle with such a ratio for its sides is believed to be harmonious and pleasing to the eye.

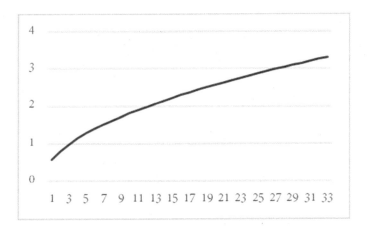

Figure 134: A graph for the ratio between the square's and the rectangle's perimeters. A value of 1 occurs at the magical number of 3.025.

The fractal root, on the other hand, is where one shape can morph into another while preserving the area and maintaining a perfect balance between the sides. Thus, the fractal root is another method that nature can implement to create self-similar fractal copies of itself, however, in dynamic geo-numerical balance.

The Cube Fractal Root

The fractal root is not restricted to square ones only; we can also extend it to include cube roots and beyond. To find the cube fractal root of a number, we start by multiplying it by 100 and then taking its principal cube root. So, for number 1, the cube root of 100 is equal to 4.6415888... Multiplying this number with its 1/10 value twice results in the number 1 back again: $4.64158... \times 4.64158... \times 4.64158... = 1$. And similar to the square fractal root, we also have two roots: the GFR and the LFR (a couple of them).

Below we list the cube-GFRs for the same constants we encountered in the previous section. Notice how the fractal cube root of the fine structure constant is almost equal to 24, a number of particular importance to the Wave Theory of Numbers and Prime Factorization, as well as the magical quartet.

Number	Value	GFR
Pi π	3.140...	6.798...
Euler e	2.718...	6.477...
Golden section Φ	1.618...	5.449...
Number 2	2	5.848...
Fine structure α	137	23.95...

Table 28: The cube fractal roots of some of the most important constants of nature.

Geometrically speaking, the fractal cube root transforms a cube into a parallelogram with fractally identical dimensions while keeping the volume the same, as shown in the figure below. It also transforms a sphere into an elongated spheroidal shape.

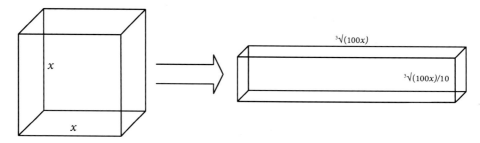

Figure 135: Using the cube fractal root to transform a cube into a parallelogram having the same volume but with fractal ratios for the sides.

Similar to the 2- dimensional case, where the perimeters are different for the two shapes, in this case, it is the areas that are different, with their ratios increasing with dimensionality, as shown below.

The ratio of the areas is equal to 1 at exactly $x = 1.36089\ldots$ And for the value of 1.37 (a fractal of α), the ratio is 1.009, very close to 1, just like π and e in the square fractal root case. (This alludes to a possible relationship between mathematical and physical constants on the one hand and their corresponding principal-to-fractal perimeters or areas ratios equal to 1 on the other.)

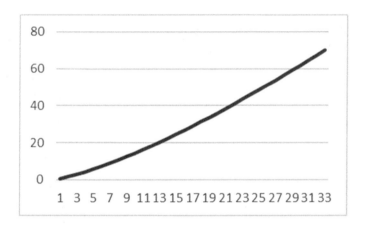

Figure 136: A graph for the ratio between the cube's and parallelogram's areas. The ratio value of 1 occurs at the magical number of 1.36.

Higher orders of fractal roots, like the fourth or fifth root, etc., can be found in a similar manner. For the 4th fractal root case, we multiply the radicand by 10 and then take its principal root. The four fractal roots will consist of one GFR and three LFRs. So, for number 16, we find $^{4fr}\sqrt{16} = \sqrt[4]{16000} = 12.649 = $ GFR (16). Consequently: $16 = 12.649\ldots \times 1.2649\ldots \times 1.2649\ldots \times 1.2649\ldots$

Therefore, to calculate the nth fractal root of any number x, we start by multiplying x with 10^{n-1} and then take its principal n^{th} root: $^n\sqrt{(x \times 10^{n-1})}$. The root we get is the GFR of x. The roots always involve a single GFR, along with $(n-1)$ LFRs. So, 1 LFR for square fractal roots, 2 LFRs for cube fractal roots, and so on.

Even though the fractal root is mainly derived from applying the principal root operation, there is still a crucial difference between the two. For the principal root, we can write it in terms of successive roots with smaller orders, something like $^4\sqrt{16} = \sqrt{(\sqrt{16})} = \pm 2$, and $2 \times 2 \times 2 \times 2 = 16$. However, doing the same for the fractal root produces something entirely different:

$$\sqrt[2fr]{\sqrt[2fr]{16}} = \sqrt[2fr]{12.649\ldots \times 1.2649\ldots} = \sqrt[2fr]{12.649\ldots} \times \sqrt[2fr]{1.2649\ldots}$$

$$\sqrt[2fr]{12.649\ldots} = 11.2468\ldots \; and \; 1.12468\ldots$$

$$\sqrt[2fr]{1.2649\ldots} = 3.556\ldots \; and \; 0.3556\ldots$$

Now, $0.3556\ldots \times 11.2468 = 4$ and $3.556\ldots \times 1.12468 = 4$, with the total equal to 16; however, the four roots are not identical anymore, neither in their number sequencing nor in their decimal point (11.2468... vs. 3.556...). Therefore, not all mathematical operations that work for the principal root will also work for the fractal one.

Generally speaking, the fractal root provides a novel approach to derive numbers from each other, whether directly, as in the LFR/GFR, or in a bit more complicated manner, as we saw above for the many fundamental constants. Additionally, square and higher fractal roots have the ability to transform geometrical shapes into ones having ratios of fractal nature, similar in principle to those based on the golden section and other fundamental constants, rendering it one mathematical operation of unique geo-numerical essence.

Part III

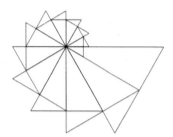

MUSIC AND SOUND

"I would teach children music, physics, and philosophy; but most importantly music,

for the patterns in music and all the arts are the keys to learning"

-Plato

In their endeavor to understand the nature of music, scientists and philosophers have attempted to formulate the experience within a numerical and mathematical context. Of the earliest attempts is that of the Greek philosopher Pythagoras. One story goes, while passing by a hammersmith shop, Pythagoras realized that specific hammers produced pleasant sounds when struck together, while others didn't. He eventually concluded that for specific weight ratios, such as 2:1, 4:3, and 3:2, the hammers produced consonance notes, while for other ratios, the notes were dissonant.

While this story's authenticity cannot be verified, the key concepts sure can; specific ratios do create pleasant notes while others don't. So, we have the octave with ratios of (2:1), the perfect 4th (4:3), the perfect 5th (3:2), etc.

Since then, music and numbers were forever bonded. Nevertheless, there were always some discrepancies between exact numerical models and the correct notes to which musicians tune their instruments. In this section, we will show that not only exact numerical models can be made to fit the correct musical tones, but geometrical shapes can as well. This will enhance our understanding of music theory, which, in return, will allow us to retune the musical scale into the correct natural Pythagorean tuning of 432 Hz.

The Geometry of Sound

"There is geometry in the Humming of the String.

There is Music in the Spacing of the Spheres."

-Pythagoras

The 432 Hz Tuning: Geometry in Music

As we explained earlier, music and math belong to two different regions of our brains. And while our eyes are continuously searching for numbers and ratios within the dimensions of the images they see, our ears are looking for the same ratios, but in the tones they hear instead.

All those mathematicians who were either professional musicians or simply enjoyed the experience elaborated on the two fields' subtle links. These links, of course, are more between numbers and music than they are between complicated mathematical functions and the latter. Special ratios and magical constants have all been expressed within compositions to create this unified experience of numbers and music, performing together in a state of harmony and wholeness. And as we will show, music is a conduit not only to evoke emotions and feelings; it can also create physical geometric patterns that are unique to the tones that produced them. By the same token, geometry creates a matrix through which the various notes can be organized, producing a powerful visual image that transforms the flat 1- dimensional musical scale into an elaborated 2- dimensional and even 3- dimensional projection.

Studying individual sound tones can be achieved by examining their physical qualities, such as frequency, wavelength, amplitude, etc. For music, however, the task is more complicated; we need to study the phenomenon as a whole, as music is not a momentary experience; it can only be perceived over time. Moreover, putting forth a mathematical model that describes some aspects of music is feasible, as we saw earlier in the Pythagorean ratios. However, formalizing the whole experience within the constraint of a few math qualities or quantities is almost an impossible task. There is so much involved in the music experience that adds more variables and dimensions to its theory.

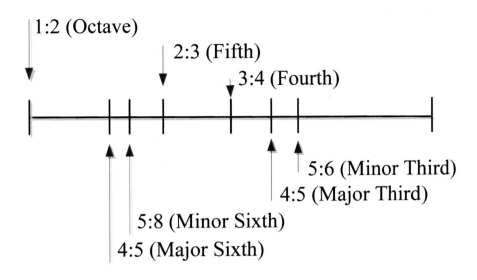

Figure 137: The various notes' ratios played on a string, along with their respective names.

We will investigate one of these dimensions by illustrating the connection musical notes have with the geometry of certain regular polygons.

One of the most effective ways to quantify musical notes is by measuring their frequencies, the rate of one complete vibration per second, measured in units of inverse second or *hertz* (Hz). These notes repeat within a harmonic ratio called *the octave* calculated by doubling or halving any Hertz value. There are a total of 12 notes to the Western musical scale, with the 13th being equivalent to the first, only an octave higher. For example, notes $A_4 = 216$ Hz, $A_5 = 432$ Hz, and $A_6 = 864$ Hz are all different octaves or pitches of the same A note. (The subscripts are references to each note's respective octave.) We used the tuning pitch of $A_5 = 432$ Hz instead of the familiar one of 440 Hz because it is an essential requirement to illustrate the connection between geometry and sound.

Below is a figure illustrating two whole octaves tuned to the 432 Hz pitch. Notice how moving up the note scale requires progressively smaller distances, like guitar frets.

Musical tones are often distributed around circles, very much like the *D*-circles, to reflect the cyclic behavior of the octave. In fact, the way the musical scale develops is very similar to a spiral or vortex, with each loop corresponding to one full octave, as shown below. Writing the musical notes in a spiraling fashion is a perfect representation as each turn of the spiral resembles the other, indicating how each octave is similar in its constitution to the previous one, however higher in pitch, which is reflected through the spiral's ascending and expanding twists. In fact, the spiral is such a perfect geo-numeric pattern able to represent so many phenomena, as we will discover later on. It is an optimum unification principle.

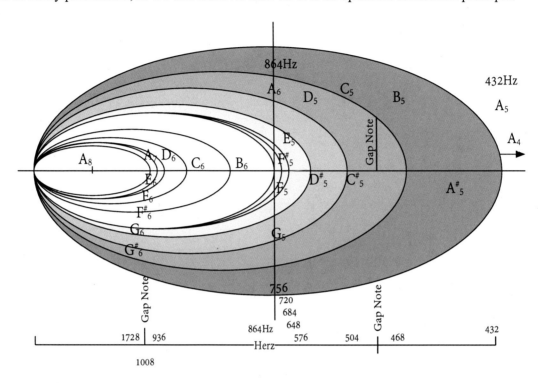

Figure 138: Two octaves and their notes. Notice how the distances between the notes' frequencies get smaller as we progress from the 5th to the 6th octave.

Connecting sound with shapes is not a new idea. It is well established that certain frequencies of sound create specific geometrical patterns, such as in Chladni figures (shown below), in reference to the German physicist and musician Ernst Chladni (1756 - 1827), who generated geometrical shapes using a violin bow drawn over a piece of metallic slab covered with sand. They can also be produced through an instrument called the *harmonograph* or even using tanks filled with water. Nowadays, these shapes are commonly known as *cymatics*.

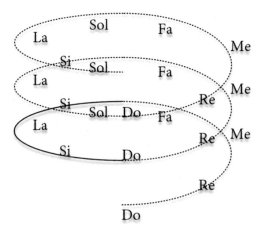

Figure 139: The octave progresses in a spiraling manner, similar to how *D*-circles do.

Producing shapes through sound may not seem very intriguing at first. After all, sound is basically a vibrational phenomenon, and when a piece of metal with sand on top of it vibrates, as in Chladni's patterns, the sand particles will also vibrate and move around, forming different shapes. However, when the vibrations form perfectly symmetric geometrical patterns, the phenomenon becomes much more interesting.

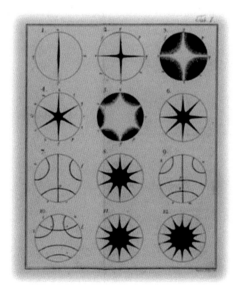

Figure 140: A sample of Chladni music-induced patterns. Note the perfect symmetry of the geometry.

Even though producing these sound-generated shapes is an easy process, putting forth a mathematical formalism capable of fully explaining the process is very difficult and complicated. This is because these shapes not only depend on the sound notes and their properties, but also on the material of the vibrating surface, the medium used to create the shapes, and the ambient temperature, among many other factors. Nevertheless, on a fundamental level, we will show that linking music to geometry can be achieved, and with a high degree of precision. We start by setting up the table below, which lists the harmonic relationships between notes on the chromatic scale and their degrees' references. The *Length* field refers to the notes' ratios for a string length of 12 units. By taking the values of the *Length* field in *Decimal* and multiplying them with 360, we get the *Degrees* reference for the notes. For example, $11/12 = 0.916$ and $0.916 \times 360 = 330°$, and so on for the rest of the notes.

Note	Hz	Decimal (Hz/432)	Length	Decimal	Degrees	Internal angle	Sum of Angles
A_5	432	1.00	12/12	1.00	360°	0°	0°
$A\#_5$	450	1.0416	11/12	0.916	330°	30°	30°
B_5	468	1.083	10/12	0.833	300°	60°	180°
C_5	504	1.166	9/12	0.750	270°	90°	360°
Gap I	-	-	-	-	-	-	540°
$C\#_5$	540	1.250	8/12	0.666	240°	120°	720°
D_5	576	1.333	7/12	0.583	225°	135°	1080°
$D\#_5$	612	1.416	6.75/12	0.562	210°	150°	1800°
E_5	648	1.500	6.48/12	0.540	195°	165°	3960°
F_5	684	1.583	6.24/12	0.520	187.5°	172.5°	8280°
$F\#_5$	720	1.666	6.18/12	0.517	186.32°	173.68°	9900°
G_5	756	1.750	6.124/12	0.510	184.76°	175.24°	13140°
$G\#_5$	792	1.833	6.062/12	0.505	182.4°	177.6°	26640°
Gap II	828	1.916	6.039/12	0.503	181.2°	178.8°	53640°
A_6	864	2.00	6/12	0.5	180°	Next Octave	0

Table 29: The harmonic relationships between notes in the chromatic scale and their degrees' references.

The *Internal Angle* field is calculated by subtracting the *Degrees* field from the starting point, being the note A_5. Therefore, for the note B_5 (468 Hz), the internal angle is $A_5°$ – $B_5°$ or $360°-300° = 60°$, which corresponds to an equilateral triangle. Moving a half-step from A_5 to $A_{\#5}$ spans a 30° angle, a shape commonly used as a symbol for the compass. The note C_5 produces an inscribed square since its internal angle of 90° is 1/4 of the 360° unit-circle. Note $C_{\#5}$ produces an inscribed hexagon since the angle is 120°.

The pattern continues through the audible spectrum of sound and is calculated for one full octave, as shown in the table below. The *Arc Length* field is calculated by measuring the angular distance around the circle from the starting point (note A_5). For example, note $C_{\#5}$ produces an inscribed hexagon since the arc length of 60° is 1/6 of the unit circle, meaning there would be a total of 6 equal sides to the polygon, and so on for the rest of the notes.

Note	Internal Angle	Arc Length	Polygon
A_5	0°	0°	Line
$A_{\#5}$	30°	30°	Angle/Compass
B_5	60°	60°	Triangle
C_5	90°	90°	Square
$C_{\#5}$	120°	60°	Hexagon
D_5	135°	45°	Octagon
$D_{\#5}$	150°	30°	12-sided
E_5	165°	15°	24-sided
F_5	172.5°	7.5°	48-sided
$F_{\#5}$	173.68°	6.32°	57-sided
G_5	175.24°	4.76°	75-sided
$G_{\#5}$	177.6°	2.4°	150-sided

Table 30: The internal angles and the arc length of the various notes and the polygons they would correspond to.

All further progressions of regular polygonal geometry follow the same logic with perfect symmetry, as shown below. Notes F_5, $F_{\#5}$, G_5, and $G_{\#5}$ correlate to polygons of 48-, 57-, 75-, and 150-sides. Therefore, they are not shown as they closely resemble a circle.

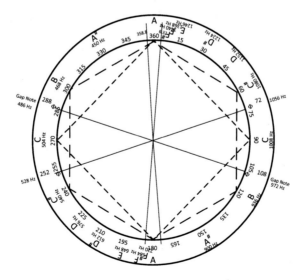

Figure 141: Note C and $C_\#$ create a square and a hexagon, respectively.

170

The F$_{\#5}$ note is of particular interest, as its length is equal to 6.18/12. The number 6.18 is equivalent to 10/Φ, where Φ is the golden section. Hence F$_{\#5}$ can be written as 10/(12×Φ) = 0.8333...Φ. The value of 6.18/12 = 0.515 is almost identical to Φ/π, and the number 0.8333... is nothing but 5/6, another hexapentakis relationship.

Worth mentioning is that the two notes used to tune the musical scale, A$_5$ and C$_5$, correspond to the circle and the square, the two opposite shapes that have the same sum of internal angles (360°). (These are the same two shapes that mathematicians and philosophers have been struggling to unify in what is known as the *squaring of the circle* problem, which we discussed earlier.)

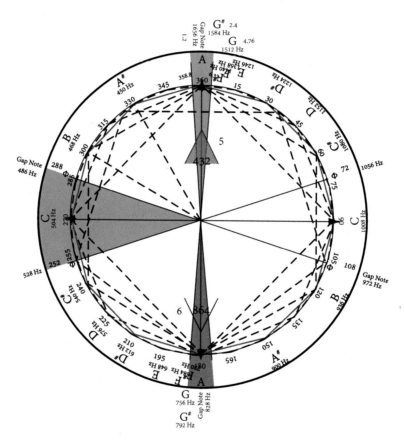

Figure 142: All notes values plotted around a circle, connecting the musical scale to arc lengths, interior angles, and natural progressions of regular polygons.

The pentagon correlates to dissonant or enharmonic gap notes like B$_{\#5}$, which are not used in the classical chromatic scale of music. Therefore, this shape is missing from the list of polygons in the above table.

(The decagon is also missing; however, we can think of a decagon as double pentagons.) Also missing from the list is the heptagon, the nonagon, and the hendecagon, which correspond to symmetries that are not observed in nature. This is of particular interest, as not only the musical notes correspond exactly to certain polygons, but also they correspond only to those polygons observed in nature and dismiss those that are not. Pentagons, however, are very much observed in nature and on many levels. Thus, maybe we should reconsider those gap notes that correspond to this geometry and reevaluate their role. They may not be that pleasant to our ears; however, they may play an important role in nature in a manner that we do not yet understand.

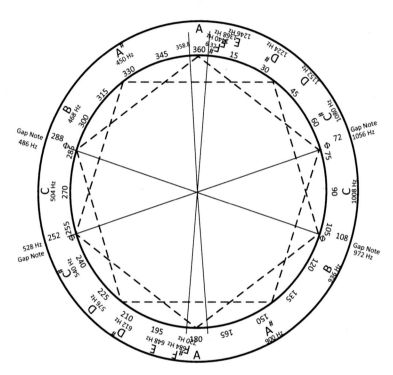

Figure 143: Two pentagons, one in reverse to the other, correlating to dissonance or gaps in the musical scale.

We were able to establish the above precise correspondence between the twelve notes of the octave and regular geometrical polygons only when we implement the pitch tuning of 432 Hz for the A_5 note instead of the modern value of 440 Hz. This fundamental number is crucial to our understanding of music, geometry, and physics as well, being the result of multiplying the central number of the wave matrix (6) with that which defined geometry (72).

All the above establishes a reconsideration of the standard pitch tuning value as the modern value of 440 Hz destroys the geometrical correspondence entirely. But tuning instruments to 432 Hz is believed to produce notes that are not very pleasant to our ears. How can we reconcile this?

In the next section, we solve the major 3rd problem and show how by a slight change of the temperament tuning, we can achieve a 432 Hz for the A_5 note and a perfect musical experience.

The Precise Temperament Tuning

Just tuning requires the Pythagorean comma (1.0136) adjustment to fix the major 3rd problem where the ratio 5/4 is not the correct ratio for this major. This is because the 5/4 ratio is wholly inadequate as a variable approach for the major 3rd, as, if it continues, it will never achieve a correct doubling of an octave, e.g., 843 Hz vs. the correct doubling value of 864 Hz. This is the interval that totally destroys Just Tuning as a viable tuning despite its clear mathematical correspondences for all other notes as well as geometrical shapes, as we showed above.

Equal temperament is also problematic because it simply cuts the octave range of frequency into 12 equal parts (based on $\sqrt{2}$) where there is a slight variation between tonal transitions exhibiting yet again wave behavior. Moreover, equal temperament loses a lot of the benefit of the base-9 mathematical system as the numbers may no longer sum to 9 in modular arithmetic.

Tuning requires correct mathematical ratios for the perfect 5th, major 3rd, and octave doubling. All the ratios of the other notes can be found within these critical ratios. The table below lists the intervals and their ratios and the difference between the Just Scale values and their corresponding Equal Temperament values. The intervals in shaded gray are the ones that need to be adjusted; however, as the 5th dissonant is inaudible, it requires a minute adjustment that is basically unnecessary.

Interval	Ratio to Fundamental Just Scale	Ratio to Fundamental Equal Temperament	Δ Ratio
Unison	1.0	1.0	+ 0.000
Minor Second	25/24 = 1.0417	1.05946	-
Major Second	9/8 = 1.1250	1.12246	-
Minor Third	6/5 = 1.200	1-18921	-
Major Third	5/4 = 1.2500	1.25992	+ 0.008
Fourth	4/3 = 1.3333	1.33483	-
Diminished Fifth	45/32 = 1.4063	1.41421	-
Fifth	3/2 = 1.5000	1.49831	- 0.001
Minor Sixth	8/5 = 1.6000	1.5874	-
Major Sixth	5/3 = 1.6667	1.68179	-
Minor Seventh	9/5 = 1.8000	1.7818	-
Major Seventh	15/8 = 1.8750	1.88775	-
Octave	2.0	2.0	+0.000

Table 31: Table of music notes and their ratios. The major third and the fifth music intervals require adjustment in Just Tuning.

A potential solution to this conundrum works by simply replacing the major 3rd ratio of 1.25 with 1.26 (derived as $2^{1/3}$ instead of 5/4); this brings the entire scale into very close equivalence with temperament tuning. This new system is called Precise Temperament Tuning.

But how does this new tuning sound? Based on the ratios and the very close equivalence to Equal Temperament values in 432 Hz, it sounds great, as early testing has revealed. Another plus is that all the new precise frequencies sum to 9 naturally, just like the sum of angles of all regular polygons/polyhedra/polytopes do. Below we list the Precise Temperament tuning values for the 5th octave.

Equal Tempered 432 Hz	Δ	Note	Δ Ratio	Precise Tempered 432 Hz	Ratio to Fundamental Precise Temp.
432	+0.081	A_5	+0.000019	432.081216	1.00
457.688	-0.459	$A_\#$	-0.01	457.2288	1.058
484.903	+1.188	B	+0.00245	486.091368	1.125
513.737	+0.742	C	+0.0014	514.4791038912	1.19
544.285	+0.137	$C_\#$	+0.00025	544.42233216	1.26
576.650	-0.541	D	- 0.0094	576.108288	1.333
610.940	0	$D_\#$	0	610.9402589451771	1.414
647.268	+0.975	E	+0.0015	648.243670902912	1.50
685.757	+0.215	F	+0.00031	685.9721385216	1.587
726.534	-0.774	$F_\#$	-0.001	725.76	1.68
769.736	+1.982	G	+0.0025	771.7186558368	1.786
815.507	+1.126	$G_\#$	+0.0014	816.63349824	1.889
864	+1.62	A_6	+0.00018	864.192432	2.00

Table 32: Equal tempered music scale based on the 432 Hz value is achievable via the 1.26 ratio value of the $C_\#$ note. Bold ratios match equal temperament while the rest maintain Just Intervals

The mathematics of this equal temperament Precise Tuning overcomes the acoustic problem of shifting the tonal center where the spectrum of overtones will suddenly fall out of Just mathematical alignment, and the music gets shifted audibly out of tune. The table below lists the √2-based Equal Temperament scale, all in the form of based 12 roots of the number 2.

The precise tuning of 1.26 has many advantages where it achieves mathematical interval perfection of the major 2nd, perfect 5th, and perfect 4th, along with mathematical perfection of the precise major 3rd interval (^{3}√2), which doubles the octave. It also enables a mathematical approach that is versatile across keys. Moreover, it integrates the geometric mod(9) number series matching all geometric angular sums, with an overtone ascribed to decimal extensions, as we showed earlier.

Equal Tempered 432 Hz	Equal Temperament √2–Based Equation	Reduced
Unison	1.00	1.00
Minor Second	$\sqrt[12/1]{2}$	$\sqrt[12]{2}$
Major Second	$\sqrt[12/2]{2}$	$\sqrt[6]{2}$
Minor Third	$\sqrt[12/3]{2}$	$\sqrt[4]{2}$
Major Third	$\sqrt[12/4]{2}$	$\sqrt[3]{2}$
Fourth	$\sqrt[12/5]{2}$	$\sqrt[12/5]{2}$
Diminished Fifth	$\sqrt[12/6]{2}$	$\sqrt{2}$
Fifth	$\sqrt[12/7]{2}$	$\sqrt[12/7]{2}$
Minor Sixth	$\sqrt[12/8]{2}$	$\sqrt[3/2]{2}$
Major Sixth	$\sqrt[12/9]{2}$	$\sqrt[4/3]{2}$
Minor Seventh	$\sqrt[12/10]{2}$	$\sqrt[6/5]{2}$
Major Seventh	$\sqrt[12/11]{2}$	$\sqrt[12/11]{2}$
Octave	2.00	2.00

Table 33: The √2-based Equal Temperament scale, all in the form of based 12 roots of the number 2.

Precise tuning allows all musical chords (both major and minor ones) to arrange neatly within tetrahedral relationships that commence with the note D and extend through the undertones of perfect 5th and major 3rd to encompass all notes. In this structure, the mathematical relationships between musical notes, including minor 2nd, major 4th, minor 6th, and 7th reveal themselves elegantly as merely inherent to the geometrical structure they naturally form (tetrahedra coalesce to form a cuboctahedron). These geometrical-musical arrangements are shown below in figure (145) .

The musical chords of perfect 5th, major 3rd, 4th, and 7th, as well as the 2nd and the 6th, all have inherent geometrical relationships in just tuning. All major and minor chords/keys are represented in the cuboctahedron structure shown below in figure (146) (the keys not listed below would appear on the other side of the structure).

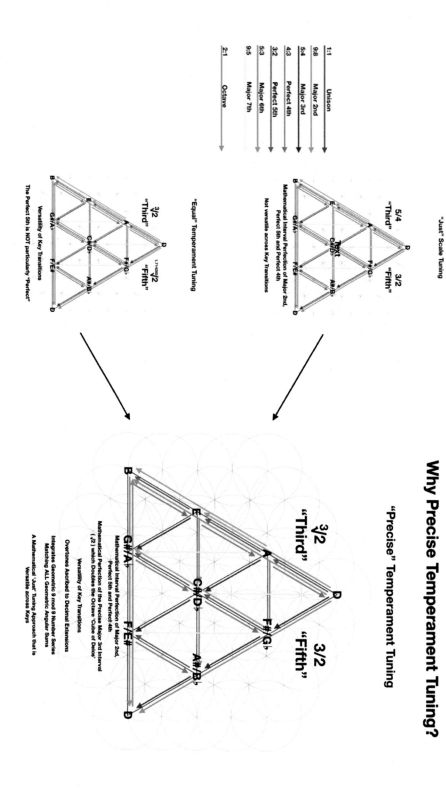

Figure 144: Geometrical precise temperament tuning embedded within a tetrahedral arrangement of the Flower of Life.

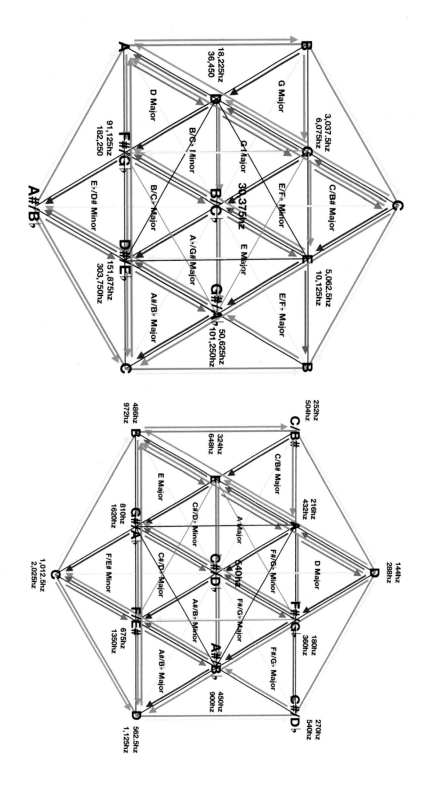

Figure 145: Left: Just Scale Tuning, a musical geometry where the cuboctahedron informs all major and minor chords. Those keys that are not shown would appear on the other side of the structure (right). Legends are the same as the previous figure.

178

The above geo-musical structure informs which transitions between keys are easily accomplished. In other words, the full understanding of the music experience cannot be accomplished without adequate knowledge of numbers and geometry. This may be what Pythagoras encrypted over two millennia ago, and it is now the time for this encryption to be broken and for his message to be deciphered.

The idea that the Flower of Life can work as a blueprint for musical chords is not surprising, given that circular waves doubling and interference can also create the 12-based configuration where these musical notes emerge naturally, as we discussed earlier in the wave theory of numbers section, reproduced below with more details.

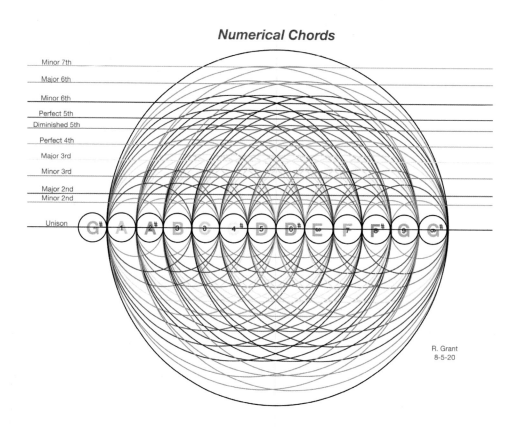

Figure 146: Numerical circular waves forming musical chords through their interference pattern.

The same number used in achieving this tuning, $\sqrt[3]{2} = 1.26$, is related to one famous mathematical problem of doubling the cube. The origin of this problem goes way back to ancient Greece, where the legend goes that the citizens of Delos, an ancient Greek city, consulted the oracle at Delphi on how to overcome a plague that hit them. The oracle replied that in order to appease the deity Apollo, who was responsible for the plague, they needed to double his altar, which was in the shape of a cube.

Therefore, given a cube having sides of 1 unit, what would its size be after doubling it? Of course, the answer is $\sqrt[3]{2}$, which allegedly came from Plato, who understood doubling the cube to mean doubling its volume. Plato, hence, advised the Delians to study math and geometry to calm down the angry Apollo.

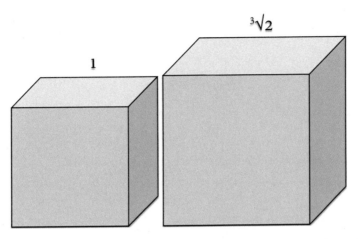

Figure 147: The Cube of Delos. Doubling the size of the cube is equivalent to doubling its volume. A cube of size 1 unit will double into a cube of size $\sqrt[3]{2}$ units.

This is an interesting story that sheds light on the important status numbers and geometry used to hold in ancient civilizations. And, as it seems, 1.26 is one such important number that does not only solve the problem of doubling the cube but doubling the octave as well.

Numbers, geometry, and music, disciplines that may look different at first; however, they are nothing but a different perspective to the same idea.

Base-12 Numeric System

"Wherever there is number, there is beauty."

-Proclus

As explained in Part I, the wave pattern replicates every multiple of 6, which doubles into 12, then 24, etc. Number 12 is unique in particular, as it can be found on all levels of the natural world. For example, we have 12 monochromatic colors, 12 notes for one octave, 12 months per year, etc. The day is made of 12 hours, as well as the night, which may have originated from the fact that there are 12 full lunations in one solar year.

Mathematically speaking, it is known that the infinite sum of all natural numbers from 1 to ∞ is equal to the unexpected value of $-1/12$, first proven by Euler. This is a very counter-intuitive result that seems to make no sense, except it is used in many important theories, such as string theory, and proven to be correct. This property does imply that number 12 has a unique status among all other numbers, at least in its $1/x$ form.

The numeric ratio of numbers 6 and 5, which correspond to the Hexapentakis configuration, is equal to 1.2, and $12\times5 = 60$, the numbers used by the Sumerians in their sexagesimal-based numeric system. Even the two most fundamental constants of π and Φ are related through this specific ratio, where $\pi/\Phi^2 = 1.2$. Therefore, number 12 is embedded right at the heart of creation. But so is number 9, which is the base of our numeric system. And we have seen how all musical tones have digital roots of 9, as well as the angles of geometrical figures, such as in the magical angle of 720°, as well as all the completion symmetries of the *D*-Space, such as in the magical quartet. So, what could the most natural base-system be, 9-based or 12-based?

Or could it be both? Could a base-12 integer system be lying hidden within the current base-9 numerical system?

In this hypothetical hybrid system, we would have the basic nine numbers, from 1 to 9, along with another three that are embedded in between them for a total of twelve. These three new numbers should be irrational, which renders them hidden from appearing in the rational number space. Just like dark matter is lying hidden within our universe of light; still, its effect can be felt by examining the movement of farthest galaxies.

The position of the first new number should be around numbers 3, 6, and 9. These are the only numbers that have irrational reciprocals that loop on themselves, creating spiral-like dynamics: 0.333..., 0.1666..., and 0.111... (The inverse of 7 creates a wave, as we have seen before.) Therefore, the optimum positioning of the first new number would be between 3 and 4, which we call *phio* (a term derived to reflect that it is right before number four). As for the second number, it is between numbers 6 and 7, and we call it *sieve* (kind of between six and seven). The third number is called *eno* (mirror image of one), coming right after number 9, as shown below.

Figure 148: The way the new numbers of phio, sieve, and eno integrate within the first nine numbers. Their new symbols are shown in the illustration: rotated 6 for phio, rotated 3 for sieve, and rotated 9 for eno.

The main reasoning behind their exact positioning comes from their values as well as their correspondence with music and digital root math. This is because planting these numbers in the specified positions creates a perfect correspondence between music, especially the major 3rd scale, and the fundamental triplet group of [1, 4, 7], [2, 5, 8], and [3, 6, 9].

From the above figure, if phio didn't exist, the major 3rd would span from note A to note D, corresponding to numbers 1 to 5. However, if we want to create a triplet match, that is for the major 3rd to start with number 1 and end with number 4 and from 4 to 7 (of the same triplet group [1, 4, 7]), two numbers should exist between 3 and 4 and between 6 and 7, respectively. This will shift the major 3rd to span from A:1 to C$_\#$:4, and from C$_\#$:4 to F:7, etc. In this hybrid system, major 3rd will correspond perfectly to all triplet groups: A$_\#$:2 – D:5, B:3 – D$_\#$:6, and so on. It is as if numbers possess overtones.

If positioning these new numbers in between the other nine was somehow straightforward, calculating their values is a bit trickier. One prerequisite property to bear in mind is that the sum of irrational phio and sieve should equal to eno, an irrational number between 9 and 10. This is a complimentary property similar to how rational numbers such as 3 and 6 add up to 9.

But where do we start looking for these numbers? Our research may provide us with some hints that can direct us. For example, the right angle triangle is of utmost importance to our thesis, and we expect these three numbers to observe this relationship, all of them or two of them at least. To ensure irrationality, we can include the square root in the calculation as it is sure to generate irrational numbers when its argument is non-square numbers, such as prime numbers, for example. Numbers 2 and 3 are perfect candidates for being the seeds for the new numbers as both are non-square numbers, and they are the basis of the binary/ trinary numbers.

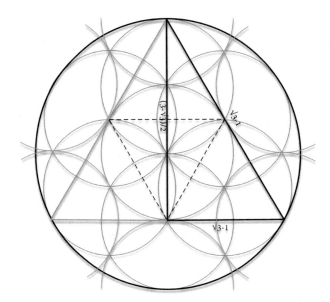

Figure 149: The Flower of Life superimposed on an equilateral triangle having dimensions that embody the values of: $(3 - \sqrt{3})/2$ and $(\sqrt{3} - 1)/2$ for phio and sieve, respectively.

As it turned out, we can fulfill all the above through one simple right triangle whose hypotenuse equal to $\sqrt{3} - 1 = 0.732...$, an irrational number. To fix the other two sides, we use a right triangle that fits inside the Flower of Life. With the hypotenuse value of √3-1, the base will equal half this value (√3 - 1)/2 and the height will equal to (3 - √3)/2. Interestingly, these two irrational values, if multiplied by 10, will satisfy all the requirements of our new numbers perfectly, with phio = (√3 - 1)×5 = 3.660254... (between 3 and 4), and sieve = (3 - √3)×5 = 6.33974... (between 6 and 7), and both are irrational, and they add up to eno = 9.999... an irrational number between 9 and 10. (And with these dimensions, the areas of the small circles of the flower will equal to π - 3, with a percentage difference of 0.04% only.)

Considering the geometrical representations of numbers from 1 to 9, those of the three new numbers, phio, sieve, and eno, can also be deduced. As shown below, the first three numbers of [1, 2, 3], coming right before phio, can be combined into one equilateral triangle inscribed within a circle, which becomes the proposed form of phio. In a similar fashion, the forms of sieve and eno can be deduced from the geometrical patterns corresponding to [4, 5, 6] and [7, 8, 9], respectively, as shown below.

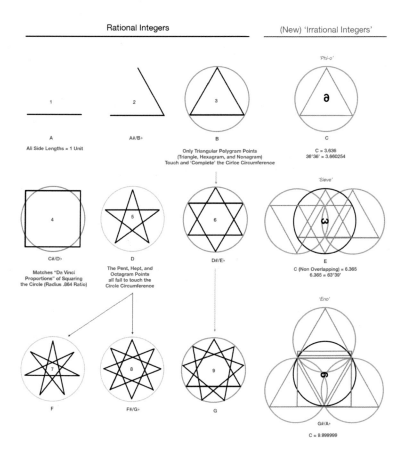

Figure 150: Progression of base-12 integer positions using ascending polygrams of equal line segments equal to 1 unit in length.

With all polygons having side-lengths of 1 unit, the first polygram forming an equilateral triangle delineates a circle (green) having a circumference value of 3.660254 (phio), after using a sexagesimal (transverse) conversion from decimal (scalar). The square that corresponds to integer 4 is too large to inscribe within the same triangle's circle. Similarly, number 5 will correspond to a hexagram that is too small to inscribe within the same circle. The same is true for 7 and 8, which are both too small to inscribe within it. Only numbers [3, 6, 9] form polygrams that fit perfectly within the original circle proportions. These nodal points of polygram inscription with the circle proceed with a 6.365 circumferential value (non-overlapping segments of the circles to form 'sieve') and 9.999… circumferential value for eno.

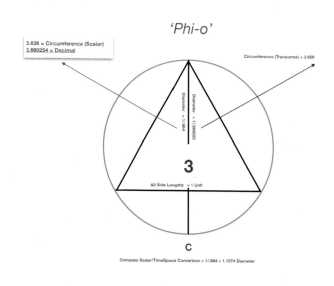

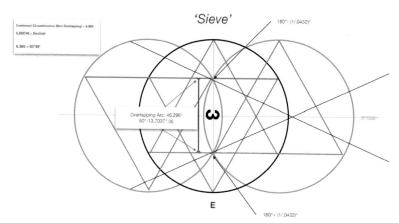

Figure 151: The geometrical representations of phio (top) and sieve (bottom) along with their dimensions values and base-10 (scalar)/base-12 (transverse) conversions.

The eno number value of 9.999... is mathematically equivalent to 10 (or 1 in digital root math). To prove this, consider the following; let's say n = eno/10 = 0.99999... (all nines to infinity). Now, we perform the following mathematical operation: $10n - n = 9n = 9.9999... - 0.9999... = 9$. Therefore we have $9n = 9$, which leads to $n = 1$. On the other hand, the digital root of eno $D(9.999...)$ is equal to 9, which is equivalent to 0 in the regular number space. Therefore, the eno number is the number that stands for the dual polarity of 1 and 0, a superposition between the two. It is the Alpha and Omega of the numbers, combined in one value.

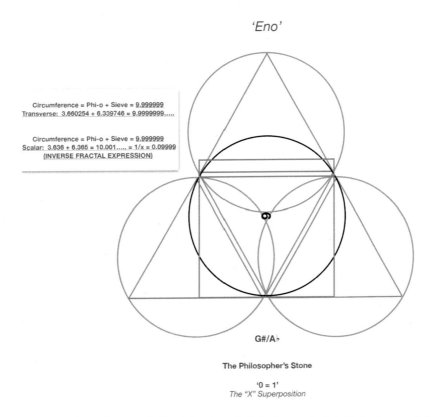

'Eno'

Circumference = Phi-o + Sieve = 9.999999
Transverse: 3.660254 + 6.339746 = 9.9999999.....

Circumference = Phi-o + Sieve = 9.999999
Scalar: 3.636 + 6.365 = 10.001..... = 1/x = 0.09999
(INVERSE FRACTAL EXPRESSION)

G#/A♭

The Philosopher's Stone

'0 = 1'
The "X" Superposition

Figure 152: The circumference of phio + sieve equals that of eno, calculated in both numeric bases.

In this sense, we can consider phio as the alpha and sieve as the omega, which is also reflected in the shapes of their unique symbols (α, ω). And due to their irrationality, they are unpolarized by definition; neither even nor odd, and not having definite digital roots either. Consequently, they won't appear in the rational number space; however, their presence is essential for the balance of the numeric scale as well as the whole universe, which will become more obvious in the next part when we couple the 12-based hybrid numeric scale with the elementary forces of nature to create one final geo-numeric matrix that unifies all.

Part IV

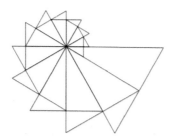

PHYSICAL SCIENCE

"Imagination is more important than knowledge. For knowledge is limited, whereas imagination embraces the entire world, stimulating progress, giving birth to evolution."

-Albert Einstein

It might be a bit difficult to understand where imagination would fit in empirical sciences. After all, science is built upon facts, hard proofs, and repeated experiments. Of course, making discoveries is an important aspect of the scientific process, but this is usually done either by expanding on an already understood model, mathematically, or experimentally, or by accident, as is usually the case. Imagination, however, is not something a scientist is expected to utilize in his research.

That being said, the reality is very different, as some of the most important scientific discoveries and breakthroughs came about through imagination, by thinking outside the mainstream orthodoxy. And there is no other field to utilize our imagination in unlocking its secrets better than physics.

Physics is the science that tries to understand nature by examining its every aspect, from the microcosm to the macrocosm. Many other fields of study that do not academically fall under the umbrella of physics, like biology and chemistry, etc. must employ physical concepts at their utmost fundamental levels. Even the DNA will have to work based on physical laws. This is why in this book, we concentrate mainly on the physical aspect of natural sciences, as by unlocking the secrets of physics, we unlock the secrets of almost all other scientific fields.

We will investigate the three main aspects of physical science: space-time, matter, and fundamental forces. We do so by employing our discoveries thus far, as well as our imagination, of course. The digital root, the wave theory of numbers, the 12-based numeric system, etc., will combine to form one complete geo-numeric picture able to explain what our universe is really made of and how it works.

And even though we will offer no experimental proof to our thesis, however, we that the numerical reasoning we propose will feel sound and logical, especially to those who appreciate expanding their horizon and freeing their minds.

SpaceTime

"The ether is not a fantastic creation of the speculative philosopher;

it is as essential to us as the air we breathe."

-J. J. Thompson

The Ether

The concept of space-time has become one of the fundamental tenets of physics, especially since Einstein's famous General Relativity theory. In this theory, the three dimensions of space, combined with a fourth one for time, create a mathematical fabric through which many of the physical phenomena, especially gravity, can have a suitable explanation. The main idea of Einstein's theory is that mass causes the spacetime fabric to curve, creating the effect of gravity, along with other consequences, such as time dilation, length contraction, etc.

One of the most important tenets of this 4- dimensional manifold is the idea that the speed of light is constant in any frame of reference, independent of the light source's motion. As counter-intuitive as this idea may be, most physicists accept it to be true. Another tenet is that space is nothing but a vacuum that contains nothing, or at least nothing important, which is how physicists deal with it today. But this wasn't always the case, as even after Einstein proposed his theory, many still believed that space is filled with the ultimate fundamental substance, the ether.

In 1887, physicists Albcrt Michelson and Edward Morley conducted an experiment that has drastically changed the way physicists view the universe. They sent two beams of light, one in the direction of the earth's motion, and the other perpendicular to it. They wanted to record any minute difference in the two beams by studying their interference pattern. There was none. This outcome of the famous Michelson-Morley interferometer experiment has been accepted by the scientific community as proof of the non-existence of the *ether*, a substance thought for thousands of years to permeate the whole universe and

189

through which light was supposed to propagate, just like waves propagate through the water. As a result, many theories that implemented the ether in their thesis were either modified or disregarded, such as the *Vortex Theory* of atoms, as the ether was deemed either non-existing or insignificant.

But what was this ether supposed to be?

The most comprehensive scientific description of the ether came from Nikola Tesla (1856 – 1943), the famous inventor, and J. J. Thomson (1856-1940), the discoverer of the electron and a Nobel Prize laureate. Tesla believed the ether is made of carriers (negative and positive) immersed in an insulating fluid and is always in a whirling motion forming "*micro helices.*" The properties of this ether vary, corresponding to movement, mass, and the existence of electromagnetic fields. In fact, Tesla believed that all the energy that exists in matter is received from the surrounding ether. Hence the ether is more like an energy repository capable of imparting its energy to all forms of matter, living and nonliving. (In today's terms, this energy is best paralleled by so-called *zero-point energy* ZPE or even dark energy.)

The ether was also used to explain the forces of nature, such as electricity and magnetism, which are disturbances through the ether, and gravity, which the pressure of the ether itself. The ether was also used to describe the elementary particles and atoms, especially through the Vortex Theory of Atoms, where the elementary particles are believed to be made of specific geometric shapes sustained in the ether through their vorticular motion.

But does the ether really exist? Or could it be that all these brilliant etheric scientists have simply got it wrong? Many scientists have come up with different experiments validating the existence of the ether, such as the Sagnac experiment performed in 1913 and the Michelson-Gale experiment performed in 1925. Many also have criticized the Michelson-Morley experiment, and others like it, as being wrongly conducted or interpreted. The famous physicist John Stewart Bell (1928-1990), who proved the existence of non-local effects in nature in the famous theory that carries his name *Bell's Theorem,* suggested that the ether may very well exist and that its existence can offer solutions to many of modern physics paradoxes, including the famous EPR (Einstein-Podolsky-Rosen) paradox. He argued that the denial of the existence of the ether was based more on philosophical grounds than on scientific ones. So why wouldn't mainstream scientists acknowledge the existence of the ether?

One part of the problem results from scientists interpreting the data coming out of experiments to suit their already held convictions. The other part has to do with ego, as it is often very hard for scientists to accept new ideas that will challenge their long-held beliefs and way of thinking. (One such belief is that Einstein's General Relativity is incompatible with the ether as a detectable substance.) Harder still is to accept the same ideas you already discredited; to accept something new is one thing, but to admit that you were wrong is another.

Nevertheless, similar concepts to the ether have slipped into the scientific mainstream, but under disguise and with many different names, such as ZPE, for example, or even under the elusive dark matter concept, which is also based on very elementary charges, similar to what Tesla and others proposed. Subatomic particles, like electrons, are thought of as *excitations of a field* by mainstream physicists. But wouldn't a vortex within the ether also be considered a kind of excitation of the ether field?

In the near future, physicists will most probably revert to the same old ether; nevertheless, to save their faces and reputations, it will all be under new names and more sophisticated scientific jargon.

The Numerical Ether

As this book is mainly about numbers and how they reside at the core of the structure of the universe, the ether being (given it exists) should be the most number-structured element. So, let us start by imagining a sea of etheric charges or carriers, as Tesla and others predicted. We numerically represent these charges with the number 9, filling all space, with each 9 having the ability to split or polarizes into numbers 3 (negative) and 6 (positive) due to the presence of charges. We are using this specific triplet group of [3, 6, 9] because it is the group that captures the essence of the ether, being a sea filled with energy and vibration, and as we showed earlier on, number 9 is the number of energy.

Imagine now an electron with a negative charge (numerically corresponding to 3) situated within this ether. What will happen is that every etheric 9 will split into 3 and 6, with the latter being the one closer to the electron. This is because numbers are attracted to those that will complete them (having a *D*-sum of 9). This split of etheric carriers will induce splits in the neighboring carriers and so on. Thus, the presence of the electron has vibrated throughout the ether and in all directions. The alignment of these split carriers matches what the lines of an electric field would look like, consequently giving the correct 1/(square-distance) dependence of the field as well.

A positive charge, a proton, with a numeric charge of 6, will have the opposite kind of etheric charge (number 3) facing it, as shown below. Consequently, the two charges will feel each other's presence via the ether's carriers and will be attracted to each other. The same situation applies to the repulsive forces between identical charges.

But why do the charges get attracted to each other to start with? The reason is due to the pressure from the insulating medium that surrounds the carriers. As Tesla, Thompson, and others suggested, these carriers create fields of force that facilitate the motion of charged bodies. After the two charges have polarized the ether in the way shown above and everything has settled, the field lines of the carriers are much denser between the two charges than on their opposite sides. Therefore, the ether pressure becomes lower in between the charges and higher in their backs, which results in a net force that punches them toward each other, as it should.

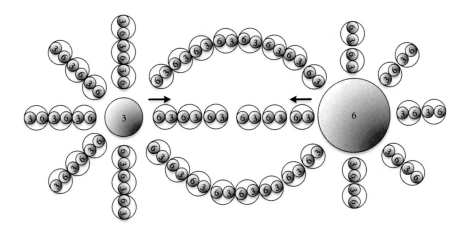

Figure 153: An electric dipole within the ether. Notice how the individual carriers of the ether will form the same exact pattern that the field lines of an electric dipole would have formed.

On the other hand, when the two charges are identical, like electron-electron, the ether charges will line up such that their field lines will be less in between, leaving unpolarized ether at the center, where the two charges face each other. This is because the ether charges will face each other in equal numeric values of 3 in this case, which is contrary to how they operate; therefore, they will start deviating from each other, as shown below. This will lead to higher pressure in the middle with the lower pressure at the edges where the field lines are denser. Therefore, the natural outcome is the charges will feel the insulating medium's pressure pushing them away from each other.

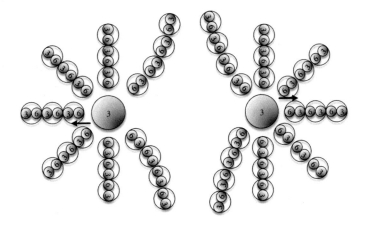

Figure 154: When two opposing charges are in the vicinity, the ether-pressure will be higher where the field lines are less dense, which is the center, pushing the charges away from each other.

192

In this sense, the ether is working as an agent by which charges can feel each other and force them to react based on their intrinsic charges. This etheric-numeric model, in my opinion, is much more realistic and physical than an invisible and imaginary electric field propagating through completely empty space. In fact, the numerical ether may surpass the electric field picture and in many more ways than by being more physical. For example, the ether model may explain one mysterious aspect of physics: the strong force inside the nuclei.

Inside the tiny nucleus of each atom exist protons and neutrons. The Standard Model of physics theorizes that in order for the highly repulsive protons to stick to each other in this tiny space, a mysteriously powerful binding force must mediate between them; the strong force. One important property of this force is that it has an extremely small range of effect, about the size of the nucleus (around 10^{-15} meters), and hence it will not be felt until the protons come very close to each other.

Using the numerical ether model, we can visualize what happens when a proton gets too close to another one, almost touching it, or maybe with a neutron in between. In this case, the protons are too close to each other for any significant ether to exist between them, as shown below. But it is through the ether that charges feel the presence of each other. Hence, if no ether exists in-between, then it follows that the insulating medium's pressure is almost nonexistent, and the pressure from the ether outside forces the protons and the neutron to remain closely packed inside the nucleus, given, of course, that the electrons' attractive force is evenly distributed around the nuclei (which is what main-stream physics already claims). Once the distances between the protons get a little bit larger to allow for the ether to flow back in, the pressure will rise again, and they will fly apart.

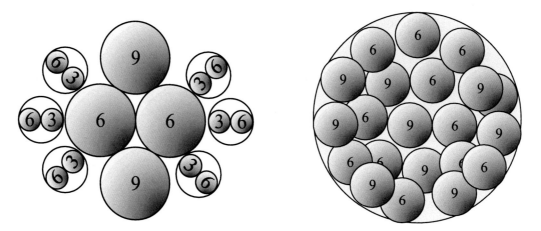

Figure 155: Inside the nucleus, the protons are too close to each other for any ether to exist between them. Hence, the pressure inside is minimum, and no repulsion exists between them.

With the number of protons and neutrons increasing, the nucleus will reach a point where its surface area becomes too big for a certain number of nucleons (protons and neutrons) to completely seal off the ether. And with these particles being in a continuous state of jiggling and vibration, the ether may start leaking in. Consequently, some protons will start to repel each other again, forcing the nuclei to disintegrate or decay through different processes depending on the specific configuration, such as proton emission, neutron emission, spontaneous fission (where the nucleus breaks into different parts), or alpha decay (where two protons and two neutrons are ejected), etc. This ejection will allow the nucleons to reconfigure their positions and seal off the ether, hence becoming stable again.

But will electrons behave in the same manner; by bringing two electrons too close to each other, will they stop feeling their charges too, and stick to each other?

Conventional physics will disagree totally, as no strong force effect has been observed between electrons or between electrons and protons. Electrons, however, are very different from protons; they are much more elementary than the protons, with the latter being almost 2,000 times heavier (precisely, it is 1,836 times) and with a diameter of 1,000 times, at least, bigger than that of the electron. Thus, if we are to think of the charges of the electrons and the protons being situated at their centers, then the distance between the charges, let's say of two protons touching each other, is around 1,000 times bigger, at least, than that of two electrons touching each other.

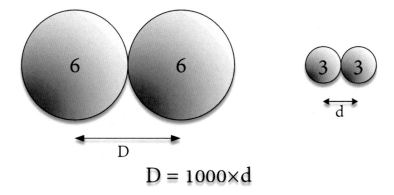

$$D = 1000 \times d$$

Figure 156: The proton is around 1000 larger than the electron and requires a million times more force to bring two electrons to the same distance as two protons.

It is already known that the electric force between charges is proportional to 1/(square distance). Hence, the repulsion force between the two touching electrons is going to be more than one million times bigger than that between the two protons, implying it is at least 1 million times harder to bring the electrons to such close proximity to seal off the ether. Let alone that the existence of the neutrons between the protons will definitely help in clearing off the ether without the need for the protons to get too close to each other,

194

a situation that does not apply to the electrons.

Thus, by proposing a numerical aspect to the ether, we were able to give alternative explanations to two of the most important phenomena of physics; electricity and the strong force, along with some of the radioactive decay processes. In fact, other fundamental forces, especially gravity and electromagnetism, may be explained through the numerical ether as well. Much work is definitely needed in order to have the correct correspondence between the numerical and the physical forces. Nevertheless, we do believe that the models we are proposing are interesting enough to be further pursued by alternate and orthodox physicist as well.

SpaceTime and the Sexagesimal Compass

The compass we use to direct us in space uses a base-60 system, divided into 360° = 60×60, along with fractions of 60 minutes and 60 seconds. Historically, we credit the Chinese with the usage of the first rudimentary compass; however, the mathematical system of its creation clearly far predates this to at least as far as our most ancient recorded civilization: the Sumerians.

Converting compass measurements into decimal is pretty straightforward. Supposing we have a measurement of x degrees, y minutes, and z seconds, the decimal value will be $x + y/60 + z/3600$. So, for a measurement of 137° 5' 6'', the decimal value is simply $137 + 5/60 + 6/3600 = 137.085$. Reversing the conversion is a bit trickier, depending on how we interpret the minutes and the seconds. For example, for a number like 117.23, there would be two different compass values. We can consider the whole number after the comma as a minute value; therefore, $0.23 = y/60 + 0/3600$, which leads to a value of $0.23×60 = 13.8$, and the whole compass conversion becomes 117° 13.8' 0''. The other option is to consider the 0.23 to be resulting from adding minute and second values. Therefore $0.23 = y/60 + z/3600$, which has the solution of $y = 13$ and $z = 48$, and the conversion value becomes 117° 12' 48''.

The compass conversion is a cyclic or circular one, which is why it is used in the longitudinal and altitudinal measurements of points on the earth, like in the GPS system, for example, as well as in calculating time, which also is measured using the sexagesimal (base-60) system. But if both time and positioning use the same sexagesimal-based system, can the compass be used to direct us through time, just like it directs us in space? Could there be a Directional Compass of time? If it does, this Directional Compass would possess the same 360° (like any circle) and has 60 minutes for each degree and 60 seconds for each minute as well; therefore, the extended clock has 360°, 21,600 minutes, and 1,296,000 seconds. These separations create demarcations of spacetime.

As shown below, the compass of time can be overlaid on a compass of direction, which reveals very fundamental and interesting points of knowledge related to our perception of both time and space. For a start, each turn of the cycle represents 15 days or approximately half a month. This is, therefore, representative of a sine/cosine relationship for each month. One day is represented by 24 hours (12 hours for the day and 12 for the night). One year is 24 and 1/3 turns of the compass: 365 days. It takes three years to complete a *meta cycle* (where there is no remaining fraction toward completion). Therefore one meta cycle of the Compass of Time is 73 cycles of 15-days each. (Incidentally, 73 is an up fractal of the fine structure constant, 0.00729735, with a reciprocal value of 137.03599). Additionally, there are 292 cycles in 12 years (or 4 completions of the meta cycles referenced above.)

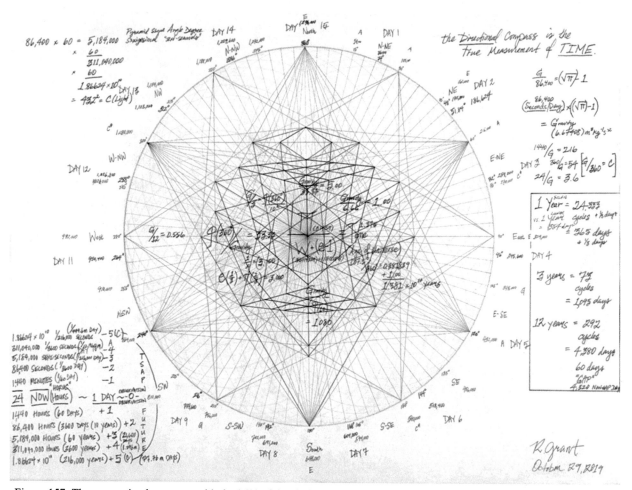

Figure 157: The sexagesimal compass, with the 360° of the circle divided into the 15 days of a half a month. The compass orients us through time just as it does through space.

The fractal patterns extend in both directions (past and future) from the moment of today. These extensions use base-60 cycles; therefore, from the above figure, fractals of time are based on multiples of 60 toward the past and future: $24 \times 60 = 1,440$ minutes; $\times 60$ (or $1/60^{th}$ of a 24-hour day) $= 86,400$ seconds ($1/3600^{th}$ of a day); $\times 60 = 5.184M$ ($1/60^{th}$ of a second and therefore $1/216,000^{th}$ of a day); $311.040M$ ($1/3600^{th}$ of a second and therefore $1/1.296M$ of a day); $\times 60 =$ ending the scale reference here at $1/777,600,000^{th}$ of a day (1.86624×10^{10} [18.6624 billion time units $= 1/216,000$th of a second]. Note that $1.86624 \times 10^{5} =$ the nearest whole number squared approximation for light speed in miles/second ($432^{2} = 186,624$). The mile as a measurement in the sexagesimal system holds its earliest references to Ancient Sumer.

197

Looking in the opposite direction now (future instead of past fractals) yields the following incredible symmetry: one day ×60 (a simple reciprocal 1/*x* value) = 60 days, ×60 = 3600 days; ×60 = 216,000 days, approximately 600 years of 360-day cycles; ×60 = 12,960 days (approximately 36000 years); and ×60 again = 777,600 days (2,160 years—the approximate length of an AEON [one zodiacal transition] on the Precession Cycle of 25,920 years).

Note the obvious symmetry of the above numbers extending both backward and forward in time scale fractals. Therefore, we can think of the "now" or "to-day" as the center or still-point with each wave expansion fractal extending as 60× or 1/60×: 0 (Now): 60× and 1/60×; 3600× and 1/3600×; 216,000× and 1/216,000×; 12,960,000× and 1/12,960,000×; and 777,600,000× and 1/777,600,000×. A full 30 days equals 2592000 seconds. This same number 25920, is the number of years in the great cycle of the precession of the equinox.

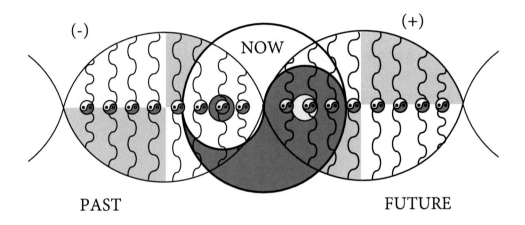

Figure 158: The past and future intertwined like waves or a double helix creating nodes where the present exist.

There appears to be an oscillation between numbers 15 and 24 (both with digital root values = 6). Number 15 (54,000 compass seconds) and 24 (86,400 compass seconds; also the number of seconds in one day, = diameter of the sun in miles/10, the same diameter of Jupiter in miles). The Great Pyramid slope angle (51.84° = 7.2^2) shows up prominently as well in regard to its compass value: 186,624 seconds = 432^2 and light speed in miles/second. Note that 54,000 is 0.625 of 86,400 and 86,400/54,000 = 1.6. Both 0.625 and 1.6 are the boundary conditions of Fibonacci (phi numbers).

All of the prominent compass positions are digital roots reducing to 9, the majority of which are prominent Hertz frequencies in Pythagorean just tuning. Note the prominent compass segmentations and the musical note chart include the following: 54, 108, 216, 324, 432, 648, 864, 5184, to list a few. These numbers are all fundamental to the compass measurement of time.

One thing is definitely clear; the compass is the perfect instrument to measure time and space. The complexity of understanding (of multiple disciplines) and the distillation down to incredibly beautiful simplicity is truly astounding. The system also clearly integrates light speed and consciousness in an interconnected web of observation. Time, in fact, is a fractal of our own consciousness.

The Golden Angle can be calculated from the following equation: $\pi/(\gamma + e - 1) = 137.5$, where γ is the Euler-Mascheroni constant equal to 0.57721... Now, 137.5°/360° = 0.381; and by adding 1 to this value = 1.381..., we get the approximate age of the universe, being 1.381×10^{10} years old. The Gravitational Constant $G = 6.67408 \times 10^{-11}$ [m^3 kg^{-1} s^{-2}] relates to all time numbers very intimately as follows: G/86,400 (seconds/day) = $(\pi^{0.5})$ - 1. Also, 1,440 (min/day)/G = 216 (there are 2,160 years in an AEON); 24 (hours/day)/G = 3.6 (360° fractal expression). Please note all the other equations where: G/3.33 = 5.00 (pentagon); G/6.66 = 1.00; G/3 = $(\Phi \times 360)/10^2$; G/(Φ/10) = 108, the list goes on and on. It appears that gravity (6.67) brings order back from disorder. Of course, all the above results from our definition of how long a second would take. But this definition is not arbitrary. In fact, it is fundamentally based on distance. If we take a pendulum of length $L = 1$ meter and swing it under the earth's gravitational field, the time it takes the pendulum to span 30° of arc is exactly 1 second. Moreover, this 30° of arc is equal to 1 cubit. This same 30° of arc, for a circle of radius of 3960 (the radius of the earth in meters), spans 7200 seconds, the same number we found in the deficient angles of all polyhedral.

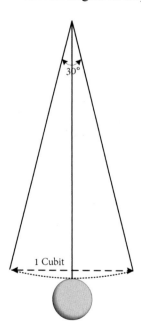

Figure 159: Each 30° of arc swing of a 1-meter long pendulum is equal to 1 cubit and takes 1 second to complete.

One interesting relationship that determines the number of seconds in a day T = 86400 (60 sec/min × 60 min/hour ×24 hours/day) is dependent on the gravitational constant G = 6.674... 10^{-11} [m^3 kg^{-1} s^{-2}] and π as follows: T = [$G/(\sqrt{\pi} - 1)$]×10^4. This is a very simple and elegant equation that depends only on gravitation. It is different from the period equation of the pendulum T = $2\pi\sqrt{(L/g)}$ and more general as the latter depends on the earth's gravity acceleration rate g = 9.81 m/s^2 while the new equation depends on the universal constant of gravitation G.

Combining the sexagesimal/compass conversions with geometry brings out a beautiful order and symmetry. Starting from the elemental geometry of a circle, a square, and a triangle, some of the major constants of nature emerge, as shown below. By setting the side of the triangle to equal π, constants emerge either in their regular or compass-conversion values, such as e = 271828; α = 729735; Ω = 567143; γ = 577215; φ = 161803; and Δ = 518436 (equivalent to the Great Pyramid's slope). Most interesting, is that the two new numbers of phio = 366025 and sieve = 63397 emerge as well as if they belong to the same league of fundamental constants.

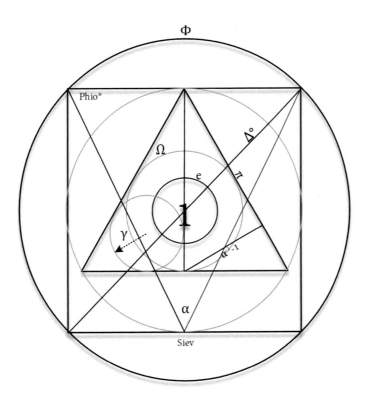

Figure 160: Using simple geometry and sexagesimal/compass conversion, all the fundamental constants of nature, along with the values of the new phio and sieve numbers, emerge from the simple geometry of the circle, square, and triangle.

The sexagesimal compass, which is a fractal of the 12-base system (60 = 5×12), works perfectly with the 9-base system to create the fundamental constants of nature and consequently define our reality. Through time and space, these two systems are intertwined, one visible, the other hidden, nevertheless, working together to keep the opposing forces of nature in an eternal balance. Whether in length, angle, or frequency, through these two bases, all is manifested, and all is united.

Matter

"Not only is the Universe stranger than we think, it is stranger than we can think."

-Werner Heisenberg

The Atom

Remember when we started talking about prime numbers, we mentioned something about the Vedic Square, the digital root of the product table of numbers from 1 to 9. We also showed how every two numbers that add up to 9 generate patterns that are mirror images of each other, as shown below.

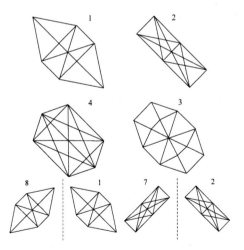

Figure 161: Shapes that correspond to numbers adding up to 9, as [2, 7], [1, 8], etc., are mirror images of each other.

Mirror imaging is a phenomenon that is observed everywhere in nature. Even on the tiniest scales of the atom, particles and their antiparticles are considered mirror images of each other, differing only through their reflected charges. Using the digital root language, we can compare the mirror symmetry of particles to our numbers and their completers, generating mirrored patterns. And when a particle and its antiparticle unite, they annihilate each other with a burst of *self-mirrored* photons or energy, which is exactly what our numbers do when added to their completers; they produce number 9, the self mirrored number. This means we can couple the four main elementary particles, the neutron (n^0), the proton (p^+), the electron (e^-), and the neutrino (v) and their anti-counterparts, to our basic numbers of the D-space, as shown below.

Particle	Anti
(neutron) n^0 $\rightarrow$ 1	n^{0-} $\rightarrow$ 8
(proton) p^+ $\rightarrow$ 2	p^- $\rightarrow$ 7
(electron) e^- $\rightarrow$ 3	e^+ $\rightarrow$ 6
(neutrino) v $\rightarrow$ 4	v^- $\rightarrow$ 5

Table 34: Coupling particles and antiparticles to the numbers of the digital root space.

We already did this in Part I when we linked the point-like aspect of particles to integers. We proved the validity of this correspondence by resorting to the most fundamental interactions of the subatomic world: the β^- decay, the β^+ decay, and the electron capture "K-capture" reaction, where our numbers worked perfectly.

- β^- decay: $n^0 \rightarrow p^+ + e^- + v^-$ and $D(1) \rightarrow D(2 + 3 + 5) = D(10) = 1$.

- β^+ decay: energy $+ p^+ \rightarrow n^0 + e^+ + v$ and $D(9 + 2) = 2 \rightarrow D(1 + 6 + 4) = D(11) = 2$.

- K-capture: energy $+ p^+ + e^- \rightarrow n^0 + v$ and $D(9 + 2 + 3) = 5 \rightarrow D(1 + 4) = 5$.

What we didn't do back then is to illustrate how fundamentally different our numeric scheme is from the one physicists usually use. We illustrate this difference through another reaction.

In the so-called "Grand Unification Theory" (GUT), the proton is expected to spontaneously decay into a positron and the neutral pion $\pi0$ (pi meson), which immediately decays into two photons as follows:

$$p^+ \rightarrow \pi^0 + e^+ \rightarrow 2 \times \gamma + e^+$$

This reaction is completely valid in the regular numerical regime, where we have a net +1 charge on the left side and a net +1 charge on the right side. So, in general, there is no problem for this reaction to take place. However, when we use our D-space numerology, we get the following surprise: $2 \rightarrow D(2 \times 9 + 6) = 6$.

Our D-space numerology is not working here. Probably because there is one catch to this reaction; it has never been observed, even after many years of research. Thus, it seems that our *D*-space numerology got it right, while that of the regular quantum numbers got it wrong (which could have saved physicists many years of research and tons of money on experiments). And there is still another advantage to our numerology.

In the regular quantum numbers, the charge numbers (-1, 0, +1) can only distinguish between negative, positive, and neutral. They do not recognize the difference between particles with opposite charges (e.g., electrons and protons) and particles and their antis (e.g., electrons and positrons), as when we add the quantum charge numbers of these particles, we get the same zero value for both cases. But these two cases are totally different; when an electron and positron combine, they annihilate each other with a burst of energy, while when an electron and a proton combine, they bond to each other, forming either a dipole or a hydrogen atom, etc.

Our *D*-space regime differentiates between these cases, as when a particle and antiparticle are combined, the result is number 9, corresponding to pure energy. On the other hand, when an electron combines with a proton, it gives 5, and they do not annihilate.

It is also interesting to see how the effect number 9 has on the first eight numbers is so similar to the effect energy has on particles, which we summarize in the following three points:

When the number 9 is added to one of the eight numbers, it raises it numerically higher, e.g., 9 + 2 = 11, which is just what energy does to particles; it raises them into higher energetic states.

Adding 9 to a number will not change the digital root of the number in the same manner energy will not change the nature of the particle; an electron stays an electron; it just gains more energy.

Adding 9 to these numbers will change their parity from even to odd and vice versa, e.g., 9 + 2 (even) = 11 (odd). Amazingly, this is very similar to what happens in atoms, as when an electron absorbs a photon, the electron will want to go up to a higher state. However, the parity of this new state must be opposite to that of the original state; odd if it is even and vice versa; otherwise, the transition is considered prohibited with zero probability. It is like saying when an electron absorbs energy, its parity is reversed. This parity reversal is not restricted to electrons only but also applies to all other particles.

Hence, it seems that the correspondence between the nine basic numbers and the subatomic world is almost exact. But how would numbers do on the atomic scale? Well, it turned out they do just as great.

Hydrogen Energy Shells: The Perfect *D*-Circle

Hydrogen is the simplest and most abundant element in nature. It is the first element to have been created and the origin of all the others, which are generated from the process of hydrogen fission with itself.

Made of a single proton and a single electron, hydrogen has always been physicists' favorite element. This is because its simplicity allowed for exact mathematical solutions for its structure, making it the perfect choice for testing new theories about the physical world, especially quantum mechanics, whose success in explaining the hydrogen spectra (the spectrum of light emitted or absorbed by hydrogen) was one of the main motives behind the adoption of this novel and bizarre theory.

The mental picture we all have about electrons being small beads orbiting protons, like the planets orbit the sun, is not that true. Electrons are more like clouds of charge vibrating with different shapes and frequencies around the nucleus and at different distances. In this cloud picture, electrons are arranged within what are called energy shells, where each shell contains a specific number of electron states. There are mainly seven shells labeled K, L, M, N, O, P, Q, depending on their angular momentum l. The number of electrons in these subshells follows the formula $2(2l+1)$, where $l = 0, 1, 2, ..., n-1$, and n is an integer, called the principal quantum number, which determines the energy of the electrons. This formula generates another 9-fold repetitive sequence in the D-space, as shown in the table below, where we have all the numbers from 1 to 9.

l (Sub– Shells)	$2(2l+1)$ (Numbers of Electrons in Sub-shell)	D
0	2	2
1	6	6
2	10	1
3	14	5
4	18	9
5	22	4
6	26	8
7	30	3
8	34	7

Table 35: The digital roots of the numbers of electrons occupying the energy subshells of the atom.

Moreover, their D-circle, shown below, exhibits many numerical symmetries, including horizontal 9-symmetry along with the 3-6-9 segmentation and D-sum.

These numbers are very peculiar not only because of their D-circle, but also because they are ordered such that they can produce a perfect spiral. Starting with a line of 5-units length, we then rotate 90° and draw a line of 1-unit length, then another 90° rotation for the 6-units, and so on until we draw the final 9-units. The emerging pattern is a simple Archimedean spiral.

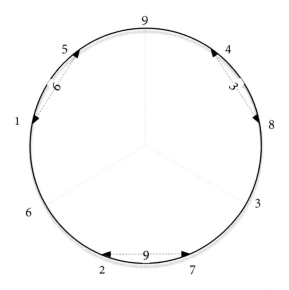

Figure 162: The *D*-circle of the hydrogen atom's energy shells.

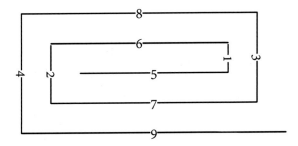

Figure 163: The numbers of the hydrogen atom's energy shells produce a perfect spiral that starts at number 5 and ends at number 9.

Could this be a coincidence, that these numbers of the hydrogen atom, the mother of all atoms, form a beautiful perfect spiral, a 2- dimensional vortex, matching what the Vortex Theory of atoms suggested years ago? Probably not, as everything we have discussed up to now supports the ether existence, even if only numerically. And if there is one truth we learned from our research, it is that numbers and the physical world are but one and the same.

206

The Periodic Wave of Elements

The natural elements comprising our universe are usually found listed in the *Periodic Table of Elements*, first introduced by the Russian chemist Dmitri Mendeleev in 1869. The table, shown below, is made of 14 -18 columns that order the elements linearly, based on their atomic number (the number of protons in the nucleus) and their chemical properties.

1 H																	2 He
3 Li	4 Be											5 B	6 C	7 N	8 O	9 F	10 Ne
11 Na	12 Mg											13 Al	14 Si	15 P	16 S	17 Cl	18 Ar
19 K	20 Ca	21 Sc	22 Ti	23 V	24 Cr	25 Mn	26 Fe	27 Co	28 Ni	29 Cu	30 Zn	31 Ga	32 Ge	33 As	34 Se	35 Br	36 Kr
37 Rb	38 Sr	39 Y	40 Zr	41 Nb	42 Mo	43 Tc	44 Ru	45 Rh	46 Pd	47 Ag	48 Cd	49 In	50 Sn	51 Sb	52 Te	53 I	54 Xe
55 Cs	56 Ba	57 -71	72 Hf	73 Ta	74 W	75 Re	76 Os	77 Ir	78 Pt	79 Au	80 Hg	81 Tl	82 Pb	83 Bi	84 Po	85 At	86 Rn
87 Fr	88 Ra	89 -103	104 Rf	105 Db	106 Sg	107 Bh	108 Hs	109 Mt	110 Ds	111 Rg	112 Cn	113 Nh	114 Fl	115 Mc	116 Lv	117 Ts	118 Og

57 La	58 Ce	59 Pr	60 Nd	61 Pm	62 Sm	63 Eu	64 Gd	65 Tb	66 Dy	67 Ho	68 Er	69 Tm	70 Yb	71 Lu
89 Ac	90 Th	91 Pa	92 U	93 Np	94 Pu	95 Am	96 Cm	97 Bk	98 Cf	99 Es	100 Fm	101 Md	102 No	103 Lr

Figure 164: The linear table of elements where the elements are distributed based on their atomic numbers.

One of the things we learned from this book is that nature is more circular than linear, with everything, including numbers, following the wave principle of fractal doubling. Naturally, this principle extends to the basic elements of matter as well. Therefore, it is logical to apply the wave doubling universal principle to the numerical references of these elements (i.e., their atomic numbers). By doing so, we find the elements being separated into their proper families in a more natural and meaningful way than they ever do through the table of elements' linear method.

As shown in the figure below, we start with hydrogen, the simplest atom, forming the first circle of elements. This circle initiates the binary doubling into two smaller ones, creating the first two complementary sine/cosine waveforms, followed by four and then eight circles.

These eight circles form another two complementary sine/cosine waves corresponding to elements He to Ne with one wave (sine) and with the other wave (cosine) corresponding to elements Ne to Ar. Due to its fractal nature, some elements will appear on both sides of the wave matrix, as when one cycle finishes, an exactly similar one starts, like an octave.

By continuing the octave doubling, elements line up perfectly in a manner that reflects their respective families, such as Metals, Lanthanoids, Acteonids, etc., with noble gases (the chemically inert members of the elements) positioning themselves on both sides of the wave matrix, at the nodes of the initial wave/circle, where everything is still and inactive. The more we move closer toward the center of each wave's peak, the more active and isotopic the elements become. Each of the elements initiates a new doubling of the waves. The carbon wave encompasses the elements of Si, Ge, Sn, and Pb, positioned at specific points or nodes that are determined by the interference pattern. The rest of the elements follows the same fractal doubling. The larger the atomic number of the element is, the farther it gets in the doubling sequence, and the more isotopes it has and the less stable it becomes.

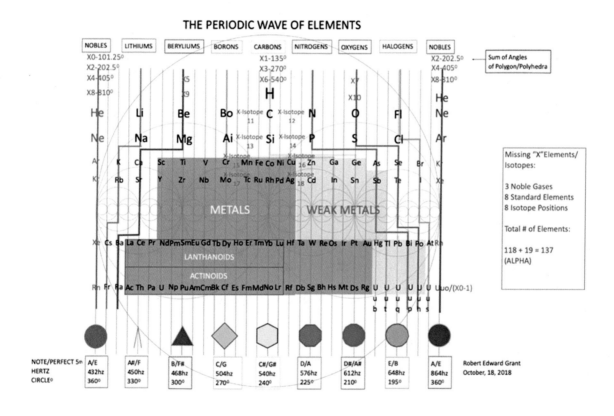

Figure 165: The wave-based table of elements along with their respective families (shaded areas) and their musical notes/geometry correspondences.

208

This wave approach allows us to associate each element with a specific frequency and shape, similar to what we did earlier on in the wave theory of numbers. For example, in this matrix, carbon (as well as hydrogen and silicon) occupy the central position, just like number 6 does in the wave matrix of numbers. Thus, carbon will correspond to note $C_\#$ of 540 Hz and also to the hexagon, which is very appropriate, as carbon chains usually take the form of hexagons in organic materials.

One unexpected conclusion of the above wave-doubling configuration concerns the existence of missing elements on the hydrogen level (or even above) such that when doubled, it will generate the next octave of the carbon level. By following this logic to its narrowing conclusion, 19 extra elements (3 noble gases, 8 standard elements, and 8 isotopes) need to be added to the already known 118 elements, for a total of 137, the same number Richard Feynman, the famous American physicist, suggested for the total number of elements (and also a well-known prime value encoding the fine-structure constant.) These missing elements may correspond to the subatomic particles as well as the basic elements of the ether. Some may correspond to the pentagon, which is missing from the above image as it correlates with gap notes, as we saw in Part III.

Whatever the nature of these suggested new elements may be, if any, the wave doubling-based configuration introduces a novel perspective to the elements and their families that will lead to a better understanding of their numerical construction as well as their geometrical and acoustic quintessence. This unified approach is another key that is getting us closer to the goal we initiated earlier on in the book: to gain a better understanding of and decipher the divine encryption.

The Forces
Gravity and Electromagnetism

"Nothing is too wonderful to be true if it be consistent with the laws of nature."

-Michael Faraday

The Basic Triplet: The Numerology of Forces

There are mainly two fundamental forces in nature, electromagnetism (EM) and gravity. In the current state of physics, these two forces are irreconcilable, being described through different mathematical formalisms that do not share any common ground. In the Standard Model, the EM force is based on the exchange of certain particles, mainly the photon, whether virtual or real. Gravitational force, on the other hand, is described by Einstein's General Relativity, which proposes a geometric solution to the origin of this force, where space and time are combined into one fabric, spacetime, that bends and twists in accordance to the masses of objects. Combining these two formalisms into one unified theory has been the holy grail of physics for more than a century but to no avail.

In this chapter, we tackle these two fundamental forces from different perspectives, resorting to numbers and symmetry only. In the process, a new understanding of the forces emerges that could shed light on how they might eventually be reconciled.

From the discussion of the last chapter, it is evident that the atomic orbitals of the hydrogen-like atoms are all about the [2, 5, 8] group. And as was mentioned before, the hydrogen atom is the origin of all atoms, not only physically but also mathematically. This is because one can approximate the wave functions of multi-electron atoms by writing them as a superposition of the wave functions of the hydrogen atom. In

210

other words, the hydrogen atom forms the basis upon which all other atoms can be built (just like prime numbers and figurate numbers are the basis from which all other numbers can be generated). So, it won't be that farfetched if we claim the [2, 5, 8] group to be the ruler of the whole microcosmic world and not just the hydrogen atom.

But what about the [1, 4, 7] group? What could it be stand for?

One hint comes from the so-called Titus-Bode law, which states that the average distances of the planets of our solar system from the sun (except for Neptune) follow an exponential curve as a function of the planets' sequence, as shown below.

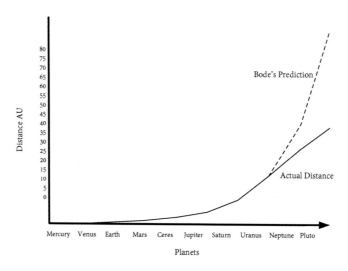

Figure 166: The above plot shows how the distances of planets from the sun follow the Titus-Bode law almost exactly and in an exponential fashion. Even though it deviates from the actual values starting from Neptune, it is still an amazing accurate approximation.

Based on this law, in order to find the distances of the planets from the sun, we first start with numbers 0 then 3. The successive numbers are then generated by doubling the ones before until we get the following sequence: [0, 3, 6, 12, 24, 48, 96, 192, 384] (all having digital roots of [3, 6, 9]). After adding 4 to all of the numbers we get: [4, 7, 10, 16, 28, 52, 100, 196, 388]. Finally, dividing these numbers by 10 gives the average distances of the planets from the sun in astronomical units (AU) (earth's average distance from the sun), e.g., 0.4 AU for mercury, 0.7 AU for Venus, and so on. This law was able to predict the orbits of Ceres (the largest asteroid in the asteroid belt) and Uranus, even before they were discovered. Amazingly, the digital roots of these orbits form a well-ordered sequence: [4, 7, 1, 7, 1, 7, 1, 7, 1] all belonging to our [1, 4, 7] group. Thus, the planets' distances follow a doubling principle, just like waves.

211

Of course, this is not solid proof that the [1, 4, 7] group is the ruler of planets in particular and the macroscopic domain in general; however, it is a hint that we will investigate further. And as we argued above, these two domains, the celestial and the atomic, are ruled by the two main forces of electricity and magnetism. Interestingly enough, it can be shown that there exists some real correspondence between the physical and numerical aspects; electromagnetism with [2, 5, 8] and gravity with [1, 4, 7]. Let us see how.

Numerology: The First Correspondence

The first correspondence between these two forces and our two groups lies in the first number of each group: 1 and 2. Physically speaking, the gravitational force is a monopole force, which means it does not have attractive and repulsive aspects similar to the negative and positive poles of electric charges or between the south and north poles of magnetism. The gravitational force is always attractive. Numerically, this monopole force will correspond to number 1, the first monopole and non-differentiated number, a number that exists in all other numbers, just like gravity affects all matter.

On the other hand, the electromagnetic force is a bipolar force (e.g., positive and negative, north and south) that embodies the bipolarity of the whole universe, which, numerically speaking, corresponds to number 2, the first polar number. Hence, these two forces' physical aspects match the numerical and mathematical characteristics of the first numbers of two triplet groups.

Mirror Symmetry: The Second Correspondence

The first thing we have noticed about the geometrical aspects of our two groups, the polygonal *D*-circles of the [1, 4, 7] and [2, 5, 8] groups, was how the former exhibited reflection or mirror symmetries while the latter did not. Interestingly, the two physical forces we are trying to relate to these two groups have the same geometrical reflection properties. This is because, in physics, one important property that distinguishes between forces is their reflection or mirror symmetry.

What this means is, if we are to reflect the spatial coordinates of a physical object (as if reflected upon a mirror) then the forces that are emanating from or subjected upon this object, whether electricity, magnetism, or gravity, etc. will either be reflected as a real physical object would do in front of a mirror, or will behave in a different manner. For example, if the coordinates of an object subjected to the force of gravity are reflected, the force lines will reflect in the same way as if they are real physical objects, as shown below.

212

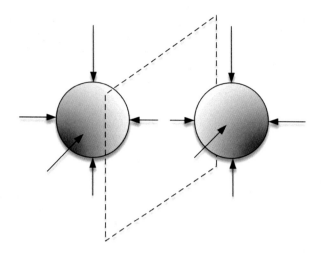

Figure 167: Gravitational force vectors behave properly under reflection

Even though the force vectors of electricity behave properly with respect to mirror symmetry, magnetism vectors don't. Take the situation shown below. When a current flows in a loop, its mirror image flows as it should; if the original current rotates counterclockwise on the right side of the mirror, it will rotate clockwise on the left side.

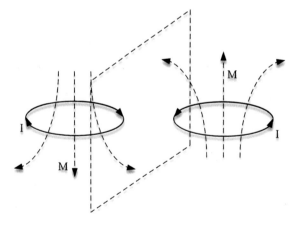

Figure 168: Even though the electric force vectors reflect properly, like gravitation force, those of magnetism don't. Therefore the whole electromagnetic phenomenon is a pseudo-vector that doesn't reflect as normal vectors do.

This circulating current produces a magnetic field that flows through the loop, as shown above to the right. However, if the mirrored loop, with the clockwise current, were to be a real one, the generated magnetic field would not point upward as if it is a reflection of a physical object; it would point downward instead, in the opposite direction, as shown above, left.

These kinds of fields or vectors that do not transform properly under mirror reflection are called pseudo-vectors, in contrast to real-vectors, like electricity. Consequently, the combination of electricity and magnetism, as in electromagnetic waves, does not observe mirror symmetry, just like the [2, 5, 8] polygonal D-circles didn't either. Consequently, the correspondence between the two triplet groups and the two forces just gets stronger.

The Poynting Vector: The Third Correspondence

When a propagating EM wave encounters a charged particle, the two forces of electromagnetism will work jointly on the particle, forcing it to move in a circular or helical fashion, depending on the sign of the charge. However, there is a third force in action here; it is the radiation pressure or the Poynting vector.

Named after the physicist John Henry Poynting (1852-1914), the Poynting vector measures the amount of directed energy per unit area an EM wave carries while propagating in space. Mathematically, it is given by the following expression $S = E \times H$, where E and H stand for the electric and magnetic fields, respectively. Notice that ($\times$) is a cross-multiplication; in a sense, we are multiplying vectors, not scalars (numbers). Hence, S itself is also a vector, and its direction is perpendicular to the plane containing the electric and magnetic vectors. In other words, S is directed in the same direction of propagation as the EM wave, as shown below.

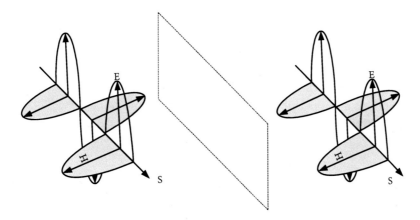

Figure 169: The Poynting vector is a real vector, meaning it reflects just like physical objects do in front of a mirror. Notice that if the wave to the left is reflected upon a mirror, the electric field will reflect properly, being a real vector. However, the magnetic field will not (otherwise, the Poynting vector would have to be also inverted). Thus the whole EM wave, as a combination, will not reflect properly either.

214

The amount of pressure this directed energy can exert is given by S/c, where c is the speed of light. This pressure is originating from the fact that the photons, the particles of light, carry momentum with them. Thus, when a flux of traveling photons hits a surface, it will exchange momentum with it, which transfers into pressure, forcing the surface to move or rotate. In fact, one idea on how spaceships may sail throughout space is based on this effect; by providing the spaceship with big sails that are pushed by the pressure of the radiation that comes from the sun. What is important to us is that we can combine both aspects, the Poynting vector and ZPE, in one theory that may explain gravity. The basic idea behind this theory rests on the fact that any object that exists in space (in other words, in the ether, which is nothing but EM fields in their ZPE states) will feel the pressure of the Poynting vector of this ZPE (just like an object is in water feels the pressure of water). This fundamental pressure is what gives the effect of gravity.

But what about our numerological correspondence? Does it make sense in this case too? Based on the above argument of the Poynting vector, E×H should give us gravity. Interestingly, we get the [1, 4, 7] group from multiplying the [2, 5, 8] with itself. This wouldn't been much interesting if it not for the fact that multiplying [1, 4, 7] with itself does not generate [2, 5, 8]; it instead generates [1, 4, 7] back again, as shown below.

×	1	4	7		×	2	5	8
1	1	4	7		2	4	1	7
4	4	7	1		5	1	7	4
7	7	1	4		8	7	4	1

Table 36: Multiplying [2, 5, 8] with itself generates [1, 4, 7]. Doing the same for [1, 4, 7] generates [1, 4, 7] back again.

In addition to the above numerical agreement, all the theories that work on unifying electromagnetism with gravity argue that gravity can be generated or modified using electromagnetism, but not the other way around, which is reflected in the fact that multiplying [1, 4, 7] with itself will not generate [2, 5, 8], but it will generate [1, 4, 7] back again. This is because, in these theories, gravity is an artifact of electromagnetism and not vice versa. This is similar to saying that water can generate pressure, but pressure by itself cannot generate water.

What could the individual numbers of each group be referring to? What does [2, 5, 8] mean for EM, and [1, 4, 7] mean for gravity? We believe they stand for different manifestations of these forces at different dimensions. Moreover, fully understanding these aspects depends on understanding the numbers themselves, which is research in the process. Nevertheless, the correspondence between the triplet groups and these forces is very interesting and definitely deserves to be carried further, which we do next.

215

Light vs. Darkness

Electromagnetism vs. Gravitation

Electricity and Dark-Electricity

As we saw in the previous chapter, many properties of electromagnetism and gravity agree with the geo-numeric aspects of the [2, 5, 8] and [1, 4, 7] triplet groups, respectively. Accepting this correspondence provides a new perspective allowing us to make novel discoveries that may not be obvious from an ortho-dox point of view.

For example, just like every number has its own complementary (adding up to 9), this principle will auto-matically apply to those concepts represented by these numbers, being an extension to the universal law of polarities, such as particles and antiparticles, odd and even, male and female, etc. As shown in the figure below, every number and its complementary create a virtual vortex, a numeric pump where energy flow in between in a manner similar to the Yin Yang concept, which stands for the unification between the oppo-site aspects of these numbers in a rotational dynamic that blends them together in circles of complete nines.

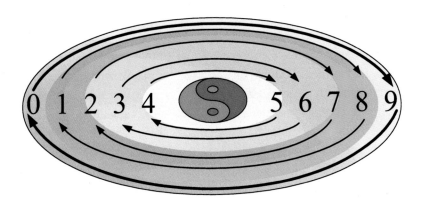

Figure 170: Each number and its complementary creates a numeric vortex of energy, flowing between the two numeric aspects.

This is similar to what we did before when we coupled the subatomic particles to numbers from 1 to 9. Those specific numeric attributions were adequate to fulfill the main atomic reactions as well as to dissat-isfy those that are not observed in nature. Alternative numeric attributions can be used as well, depending on the context in which they are used. Moreover, as we have learned from the music part, three new numbers, phio, sieve, and eno, are hidden among the original nine, completing the set to 12. From these 12 numbers, we can create a new model for the forces and the particles that represent them. In this model, some of the 12 numbers will stand for forces or particles we are already familiar with, like the electron, positron, EM, etc., while the rest of the numbers will stand for concepts that are new to us. As shown below, the 12 numbers are coupled to the fundamental particles and forces, including some new ones, as well as with the musical notes of the octave.

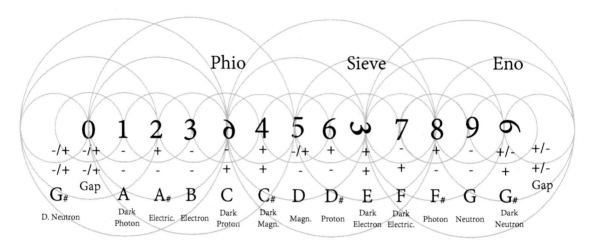

Figure 171: The 12-base numeric system couples music notes, forces, and charge/particles to their complementary aspects.

In this numeric model, the main force is electricity. Static electricity is the core property of subatomic particles, like the electron and the proton. Magnetism and radiation require movement of the charges to manifest themselves. On the other hand, static electricity is always there, even if the particle is dead-still in its place; we can never deprive a charge of its electric field. The numeric value of this force is 2, and 7 for its complementary, dark electricity, which represents the aspect of this force that is hidden for us. It is the inward radiation force of gravity that binds everything together. The particle aspects of these forces are the photon, the particle of radiation and light, and the dark photon, the particle of gravity, the graviton. From these two main forces, all other aspects emanate, such as magnetism and radiation and their dark aspects.

We list the twelve physical aspects in the table below, along with their numeric and charge-polarity values, which indicate their dark/light aspects.

Light and darkness create the boundary of what we call reality. Think of the color of a rose; what you experience as red is only that fraction of white light reflected back into your eyes; all other colors are absorbed within the texture of the petals, hidden from our sight. An entity inside the petal, let say a germ, will look at the surface of the petal and see it made of a different color, probably all colors but red. So,

217

what could the real color of the rose be, red or everything else? Obviously, it is a matter of perspective. Light is just that fraction of energy that we can see; all remaining fractions are what we call darkness, existing all around us. Light is the radiative force, spreading outwardly from its source. On the other hand, the vacuum is the opposite force, radiating inward, pulling everything to the center. It is the force of gravity. Therefore, the dark photon could be the graviton, the hypothetical particle responsible for the masses' attractive forces.

Plus Aspect	Value	Charge	Negative Aspect	Value	Charge
Photon/Light	8	-	Dark Photon/Graviton	1	-
Electricity	2	+	Dark Electricity/ Gravity	7	-
Magnetism	5	+	Dark Magnetism	4	-
Electron	3	-	Proton	6	+
Neutron	9	+/-	Anti Neutron	9	+/-
Anti Proton	Phio	+	Positron	Sieve	+

Table 37: Numbers and their associated physical aspects, along with their respective charges.

The numbers of sieve and phio stand for the antiproton and the positron, both irrational and hidden from us in most of the universe. Nevertheless, they do appear at specific places of the universe, such as at the center of the galaxies where the vorticular black holes manifest, which explains their irrational numeric values.

This numerical correspondence allows us to understand the dynamic between the different particles/forces combinations by simply following the algebraic rules of the numbers to which they correspond. For example, the electron, being number 3, can be found from adding graviton to electricity. Magnetism, 5, is 2+3, which is electricity + an electron. Electricity will force the electron to move, and a moving charge will create a magnetic field. Magnetism (5) + electricity (2) = 7, the graviton, which agrees with our previous discussion of the Poynting vector. Magnetism (5) + electron (2) = photon (8). And so on,

Similar to sound, colors also have their share of numeric representation. As shown below, infrared will correspond to number 1; red is number 2; orange number 3; and so on, all the way to ultraviolet, the number 9.

The new numbers of phio, sieve, and eno would correspond to hidden frequencies of sound and light. Even though we may not hear nor see them, they are still as profound as the rest of the frequencies, similar to ultraviolet or infrared, etc.

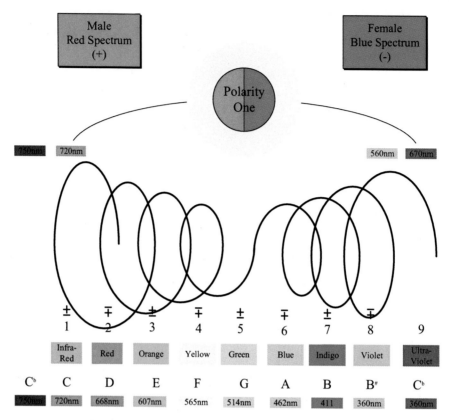

Figure 172: The correspondence between the main numbers and the colors of the spectrum.

Transverse vs. Longitudinal: Light vs. Sound

One important aspect missing from the above correspondence is sound. Sound is another form of waves, called longitudinal waves, which are different from the transverse waves of light as they oscillate in the same direction of their propagation, whereas light waves oscillate perpendicular to their propagation's direction. And while light can travel through the vacuum (or the ether), sound is supposed to travel only through a non-vacuum medium, like air. (Nevertheless, Tesla claimed it could travel through the ether and with the speed of light.)

Transverse waves, such as light, radiation, energy, etc. form geometrical sinusoidal waves/circles that are mainly governed by the π constant. Longitudinal (compression) waves, such as sound-gravitation waves, form rectangular geometrical shapes that are governed by the Euler constant (as we showed earlier when we compared the volume of the sphere to that of the cube).

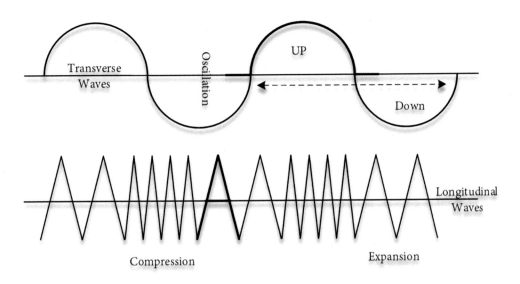

Figure 173: Transverse waves (top) vs. longitudinal waves (bottom). Transverse waves are spherical disturbances of the medium (looking like the Greek letter Omega [Ω]) that oscillate perpendicular to their direction of propagation. Longitudinal waves are triangular compression disturbances (looking like the letter alpha [A]) that oscillate parallel to their propagation.

The sound phenomenon is represented through the gravitation aspect of the numeric correspondence. Combining these two aspects together is not arbitrary action. On the contrary, recently, physicists discovered that sound waves carry mass, and they also produce their own gravitational waves. Remember that Tesla believed sound to travel through the ether with the speed of light. Therefore, sound, or its particle, the phonon, could very well be the graviton. Of course, it all depends on the frequency, and for certain frequencies, sound is able to affect the ether to produce gravitational effects.

Just like all aspects carry numerical signatures, these two forms of waves of light and sound carry numerical signatures as well, being the septenary numbers of [1, 2, 4, 8, 7, 6] (the wave-like reciprocal of 7), distributed over their crusts and turfs.

These six numbers are unique not only because they emerge from the wavy essence of 1/7, but also because they emerge from the doubling process starting from 1, all the way to 2, 4, 8, 16 (7), and 32 (5), and back to 1 again, as shown below. The [3, 6, 9] are outside the group as doubling 3 results in 6, and doubling 6 brings 3 back again. They form a close loop around themselves, standing for the particle aspect of the numbers, from electron to proton to electron, ad infinitum. Number 9, the neutron, doubles back on itself.

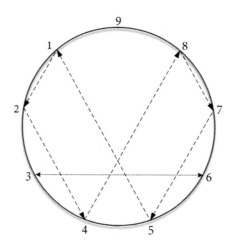

Figure 174: The doubling process of the electromagnetic pump produces numbers [1, 2, 4, 8, 7, 5], leaving [3, 6, 9] behind.

Notice how numbers 1 and 8, which stand for graviton/sound (dark photon) and light (photon), are at the top of the circle, on both sides of the neutral number 9. The process initiates from number 2, the polar number of electricity, positive and negative, male and female, which then creates magnetism (4), then light (8), then dark electricity (7), dark magnetism (5), and finally gravity (1).

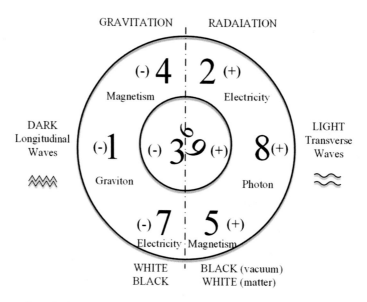

Figure 175: The two aspects of gravitation and radiation and their corresponding numerical values. At the boundary, the [3, 6, 9] group belongs to both forms of energies (here depicted to resemble the Om symbol).

221

The two forms of waves travel together, as shown below, with their corresponding numeric values completing each other. Sound scalar waves travel concealed from us, right underneath the radiation of light. At the same time, charges progress forward in their path, creating spherical waves of electromagnetic radiations and light, rectangular sound waves of gravity propagate along the same path. These sound waves, once confronted with a mass, impart their energy in the form of gravitation. The conversion factor between the two types of waves as they propagate simultaneously through the vacuum is the speed of light. In mediums such as dry air, however, they are related via the golden section constant through the following formula: $C/\sqrt{[2(1 - \varphi) \times 10^{12}]} = 343.22...$, where C is the speed of light 3×10^8 [m/s] and φ is $0.618...$

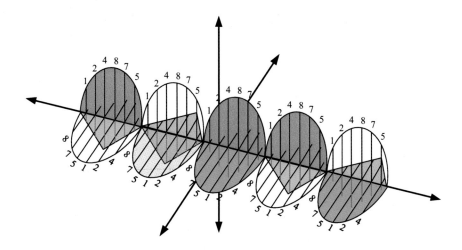

Figure 176: A propagating electron and/or proton create sound waves along with radiation waves.

Once they reach a solid medium, they act oppositely depending on the medium of propagation, and thus they possess varying speed dynamics where light, ostensibly, travels faster through the vacuum, in contrast to sound, which travels faster through denser media like granite (6,000 m/s), much faster than it does through 20° C of dry air (343 m/s). They are \opposite manifestations but for symmetrical reasons, each to their wave nature and parallel or perpendicular positioning vs. the wave direction. The way these waves reconnect results in the separation manifestations of light and dark (gravity) within the medium through which they propagate.

Thus, through the magic of numbers, we were able to create a model for the physical universe where everything is connected and completed. More importantly, it allowed us to give a reasonable explanation of what gravity might be. We know that this is not the whole story, and there are so many other factors going into the real experience of these forces. Nevertheless, what we did is to initialize a new thinking process that may lead to a much better understanding of the physical reality than the current models are offering.

And with this, our numeric journey into the scientific domain has concluded, for now at least. Many more discoveries are being made constantly, but we felt this would be enough for an introductory book that will explain the logic behind our endeavor towards better learning and understanding. And to properly end the book, we conclude with a small journey through time to illustrate the correlation our ideas have with ancient knowledge and to emphasize that what we are doing here is nothing but a smooth continuation, with a slight break up, of what has always been the aim of the philosopher; to seek the true meaning behind the divine encryption.

Part V

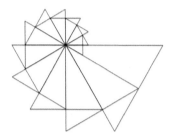

ART AND HISTORY

"Study the science of art. Study the art of science. Develop your senses,

learn how to see. Realize that everything connects you to everything else."

-Leonardo da Vinci.

Most of the material presented in this book is probably new to the reader. But how new is this approach really? Throughout its long history, humanity always looked upon the universe through a holistic lens. All the scientists and philosophers of the ancient world were considered polymaths, combining a vast amount of information from various fields of science. This, they believed, was the only way for man to attain understanding and wisdom. Spending a lifetime studying one narrow area of a very specific scientific field would have been considered futile, if not foolish. All the visionaries of the renaissance and beyond were also polymaths. Newton, Da Vinci, Galileo, Tesla, etc., all knew very well that the path of understanding should pass through all branches of knowledge, including numbers, math, geometry, music, etc. Their methods and achievements are a reminder for us that it is through holistic knowledge the most important discoveries and achievements are made.

In this section, we concentrate on one polymath figure who is known for his scientific as well as artistic supremacy. Leonardo Da Vinci's interests span many disciplines, including drawing, sculpting, architecture, engineering, anatomy, etc. He personifies the real meaning behind the word *polymath*. He was also one of the best cryptographers, embedding his drawings with many encrypted messages, some of which have already been deciphered while the rest is still waiting, and strangely, could be connected to the Great Pyramid. The Great Pyramid of Giza has withstood space and time for far more than 5000 years (if not much longer). This perplexing edifice has been a constant reminder that the past is not what we would like it to be; a steady development from hunter-gatherers to farmers, to the first cities of Sumer, and so on. Something much bigger was there, an ancient civilization with knowledge far ahead of anything we believed could have existed, even of our own. They probably left us this building, among many, to remind us of their existence as well as to tell us about their knowledge, something we are in much need of today. We touch briefly on this monument to shed light on a recent enigmatic discovery that will not only alter our perspective of the monument's history and its builders but also help us bridge the multiple gaps within this history and to connect the past with the present, as well as with the future. We conclude this part with some hand-drawn geometrical shapes by Robert E. Grant. Art is one conduit to activate the right brain and to bring harmony to the soul. A true philosopher needs to balance both science and art, right and left, Yin and Yang, etc., in order to achieve true understanding. Thus concluding this part, and mostly this book, with art creates the perfect opposite to its beginning, which was about numbers and math, uniting brain and heart in a strong bond that fulfills their destiny and brings them closer to the will of the one.

Leonardo and the Vitruvian Man
Encryption in Dimensions

"Simplicity is the ultimate sophistication."

- Leonardo da Vinci

How much information can one image conceal?

We are all used to secrets being hidden inside images, especially in old paintings, as painters often had subliminal messages they would like to include in their art. Most of the time, innocent numbers like the golden section, or even polygons like the pentagon, can be deduced from a careful measurement of the dimensions of the paintings. In other times, however, not so innocent hints are embedded within the many elements of the painting; hints related to the painter, or his patronage, belief in things that are forbidden by the authority or not accepted by the common people, e.g., heretic beliefs, political opinion, or even sacred knowledge that is not intended to be divulged to the public, not yet at least.

But how about hiding most of the above, if not much more, in a single image painted on one page of a small notebook? This is exactly what Leonardo da Vinci did in his famous Vitruvian Man sketch, shown below.

Drawn around 1490 AD, it is widely believed that the Vitruvian Man depicts a man with the perfect proportions, as Leonardo would have envisioned him. The man is enclosed within a square and a circle that seems to define the proportion of his various elements, especially arm and leg span. Additionally, along the whole torso of the man and on his limbs, we see straight lines that cut the various parts into the supposed perfect proportions intended by Leonardo.

But is this all Leonardo, the polymath, had intended to say, or is there more, much more, hidden beneath the ostensibly simple sketch? Let us find out.

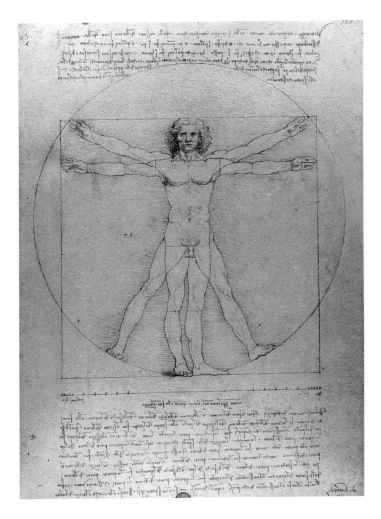

Figure 177: The Vitruvian Man sketch, drawn in one of Leonardo's many notebooks, around 1490 AD.

The Pyramid in Man

Around 1490 AD, the Great Pyramid had stood on top of the Giza plateau for more than 5000 years (some may claim it to be 10000 years, if not more). Nevertheless, at the time Leonardo drew his man, it is supposed that not much was known about the Great Pyramid, such as the precise dimensions of its exterior design, let alone its interior configuration. That being said, the idea that the exact design of this pyramid is laid down, in perfect details, within the Vitruvian Man would be not only unexpected but also unacceptable by orthodox historians. Still, it is there, in its precise and perfect proportions, as shown below.

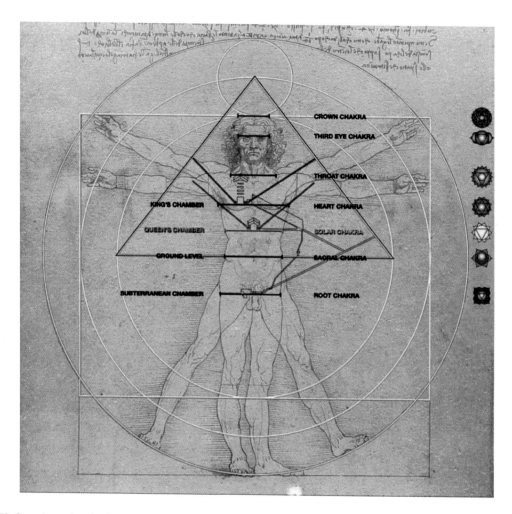

Figure 178: Superimposing the Great Pyramid design on top of the Vitruvian Man reveals an undoubted match between the various elements of the pyramid and the horizontal cut-lines Leonardo drew all over the figure.

The cut-lines we see all over the body define the precise positions of the various elements of the Great Pyramid's interior. From the subterranean chamber all the way up to the Queen's Chamber, the King's Chamber, and even the top part of the Relieving Chambers. The four shafts that extend from the King and Queen Chambers are indicated through the intersection point the cut-lines make with the horizontal arms. The naval line corresponds exactly with the base of the pyramid.

This leaves us with three horizontal lines in the drawing that do not seem to correspond to any yet known feature inside the pyramids. Could there be anything there?

These seven cut-lines are also carefully positioned to match the seven chakras, the energy portals that are believed to control the inward-outward flow of the body's energy. Is this only a coincidence, or was Leonardo hinting at something very profound about the Great Pyramid, a monument that stands for man himself?

In order to answer this question, we probably need to know how Leonardo knew about the precise design of the Great Pyramid in the first place. Where did he get the blueprint from?

Could he get it himself?

Leonardo in Egypt!

Most of the major events of the life of Leonardo da Vinci are well known and recorded. Almost all of them, as there is a period of three years where he seems to have disappeared with no record of what happened to him during this time. These are the years between 1482 and 1486.

However, there is one subtle clue to what might have happened during these missing years found in a specific letter in one of Leonardo's notebooks and documented in a book titled the *Codice Atlantico*, which is a twelve-volume encyclopedia of Leonardo's notebooks and drawings gathered in the 16th century by a sculptor called Pompeo Leoni. Of interest to our investigation lies in chapter XXI, where we find a letter that gives an account of his secret travels to Egypt working as an engineer for the Mamluk sultan "Keit Bey". In fact, historical records tell us that during this specific time, namely from 1468 to 1496, Egypt was ruled by the Mamluk sultan Sayf ad-Din Qaitbay. The two names almost match except for the letter K, which is expected, as the letter Q refers to the Arabic letter (ق) spelled Qaf, which Latin speakers cannot pronounce correctly, so they spell it as K instead. The letter Q is sometimes used to indicate that this is not exactly the K but Qaf.

In this letter, Leonardo records his travel events where he mentions going to a place called Babylon, not the one on the Euphrates in modern-day Iraq, but the one in Egypt. This is also correct, as there is a fortress town with such a name dating back to the time of the Roman occupation of Egypt.

The title of the letter is to "The Devatdar of Syria, Lieutenant of the Sacred Sultan of Babylon." In this letter, Leonardo frequently mentions one Taurus Mountain that is made of limestones. Could this be encrypted hints of the Great Pyramid? The word Taurus could be referring to the Apis Bull, the sacred bull of the Egyptian. Its hieroglyphic symbol depicts a bull along with a chevron symbol. The same chevron shape can be found at the entrance of the Great Pyramid, as shown below.

Figure 179: Left: The stones that protect the entrance of the Great Pyramid from collapsing have a chevron-like shape. Right: The same chevron shape appears in the hieroglyphic symbol of the sacred Apis Bull.

In another enigmatic text, Leonardo speaks about entering a dark place, where he had to bend first into an arch, resting on his hands and knees, where he believed some marvelous things could be discovered. Below is the relevant extract from the *Codec*:

> *"Unable to resist my eager desire and wanting to see the great ... of the various and strange shapes made by formative nature, and having wandered some distance among gloomy rocks, I came to the entrance of a great cavern, in front of which I stood some time, astonished and unaware of such a thing. Bending my back into an arch I rested my left hand on my knee and held my right hand over my down-cast and contracted eye brows: often bending first one way and then the other, to see whether I could discover anything inside, and this being forbidden by the deep darkness within, and after having remained there some time, two contrary emotions arose in me, fear and desire-fear of the threatening dark cavern, desire to see whether there were any marvelous thing within it ..."*

Could this be a description of his attempt to enter the pyramid? We may never be sure, as Leonardo was a master of encryption. Nevertheless, the possibility of Leonardo being in Egypt around 1486 is an indicator of his fascination with this land. And whether Leonardo performed this trip or not, the fact remains, the Great Pyramid is definitely encrypted within the Vitruvian Man, a strong testimony not only for Leonardo's craftsmanship but also for the important place this great edifice held at his heart, in particular, and at the heart of the whole renaissance in general.

231

The Great Pyramid: The Alpha and Omega

Alpha and Omega are the first and last letters of the Greek alphabet. They were used extensively, especially in Christian art, as a symbol for God or Jesus Christ. Its origin is traced back to the Book of Revelations, by John of Patmos, where Jesus is supposed to have said, "I am the Alpha and the Omega." In Islam, two of the main names of God are the First (Al-Awal) and the Last (Al-Akhir), Quran: (57:3). These two letters were used extensively in the Byzantine empire, where they were often accompanied by the letters Chi and Ro, which were thought to stand for the name of Jesus Christ in Greek.

Figure 180: The Chi-Ro symbol of the Byzantine Empire is believed to stand for the name of Christ.

Numerically speaking, the two letters correspond to numbers 1 and 24 (as there are 24 letters in the Greek alphabet). However, in the Greek gematria method, the omega letter is assigned the number 800. In the digital root sense, it is either 1 and 6 for the first method or 1 and 8 for the second, where they complete each other to 9.

We saw this symbol back in the geometry part, where we compared them to the triangle and the circle, the first and last polygons. We also saw them in the shapes of the transverse and longitudinal wave patterns. However, to find these two specific letters engraved on the sarcophagus inside the King Chamber of the Great Pyramid, which dates to at least 5000 years ago, long before the time of Christ, is something of great interest.

Given the huge amount of study this object has been subjected to, it is amazing it was't noticed until 2018, when it was discovered by Robert Grant, one of the authors of this book. As shown below, the two letters are clearly visible, a little bit off the scale when compared to each other (photo by Robert Grant).

Figure 181: Capturing the Alpha-Omega engraving at the backside of the Great Pyramid's sarcophagus.

When inspected in greater detail, the engraving provides a myriad of fundamental numerology, including numbers 5 and 6, 33, and the golden section, as shown below.

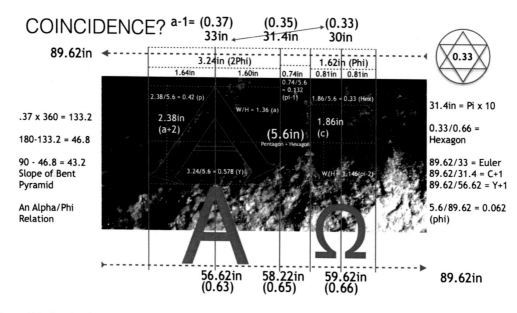

Figure 182: Conducting a thorough measurement of the A-Ω symbol reveals many interesting and symbolic numbers.

233

What could be the reason behind this enigmatic engraving? They are definitely not intended to be Greek letters, as the Great pyramid dates much older than the Greek civilization. Were they engraved later on? If so, when and by whom? Could this lead tell us something about the function of the sarcophagus? Could Leonardo have seen this engraving on his supposed trip to Egypt?

This definitely adds another layer to the enigma surrounding the Great Pyramid, as where most of its secrets are codified in numbers, dimension, orientations, etc., this is the only written inscription found inside it. (There are some hieroglyphics found inside the relief chambers on top of the King's Chamber. But these are most probably forgeries made 150 years ago.) Whatever the reason may be, finding these engravings, after all these years, is definitely one of the most interesting archeological finds of this decade.

Squaring the Circle

In Leonardo's sketch of the Vitruvian Man, he encloses the figurine within a square and a circle. However, he positions these two polygons in a way that looks a bit odd. For example, he doesn't center them around a mutual point, but rather they are off by the distance from the groin area (the center of the square) to the naval (the center of the circle). This odd arrangement could be an attempt to set the dimensions of the perfect man within geometrical limits. But as is always the case with Leonardo, there is much more than meets the eye.

The square and the circle invite the eternal conundrum of the ancient mathematicians, the *squaring the circle* problem. This problem has two ways of reasoning. One is to try to find a square that has the same area as a circle (or vice versa); the other is to find a square that has the same perimeter as a circle (or vice versa). Of course, there are no rational solutions to this problem, in either way, due to the involvement of the transcendental number π in both the area and perimeter of the circle.

But Leonardo doesn't seem to have been considering squaring the circle, not in the same line of thinking, at least. The square and the circle could be hints to something more; extra hidden geometry that can be revealed through these two obvious polygons.

For example, a pentagon inscribed within the circle will have its base just touching the upper parts of the feet of the man, as shown below.

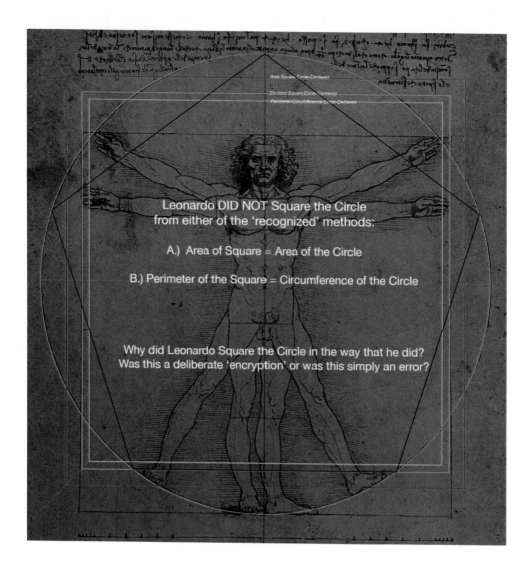

Figure 183: When placed inside the circle of the Vitruvian Man, a pentagon's base would touch the tips of the toes in what looks like an intended action.

Drawing another circle that encircles the square while touching its four corners creates the perfect boundaries for a hexagon such that the first circle drawn by Leonardo will work as the hexagon's inscribed circle, as shown below. Therefore, the original circle and square are positioned in this odd configuration to create the perfect limits for a hidden pentagon and hexagon.

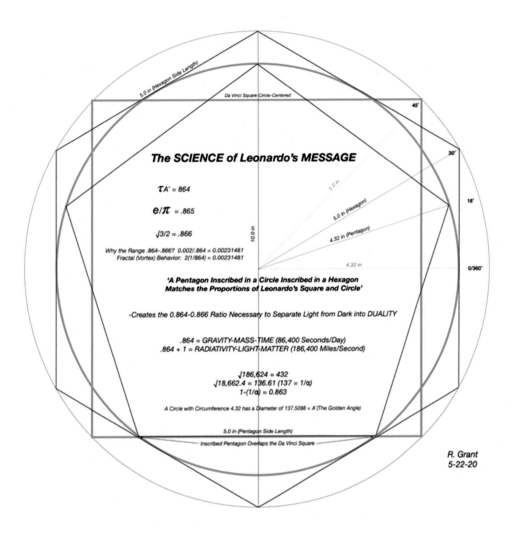

Figure 184: The hexagon and the pentagon both can be drawn within the configuration of the various geometric elements of the Vitruvian Man, especially the square and circle.

Within this geometry, many interesting numbers pop out. First of all, the radius of Leonardo's original circle is 4.32 inches, a number whose importance we are all familiar with by now. The diameter of the larger circle is 10 inches (unity/tetractis), while the diameter of the smaller circle is 4.32×2 = 8.64, a very important number that is related to the speed of light (equal to 186400 miles/second), as well as being very close to the ratio of *e* to π. And as we mentioned before, a circle with a circumference of 432 units has a diameter of 137.5, the golden angle.

236

The overall combination of these geometries in the Vitruvian Man generates an icosahedron, as shown below. A member of the five Platonic solids, the icosahedron was associated with water, thus life. (Interestingly, water molecules do cluster in icosahedral configurations.) The man in Leonardo's sketch can then be thought of as the water bearer, the symbol for the Aquarius constellation, the new coming age.

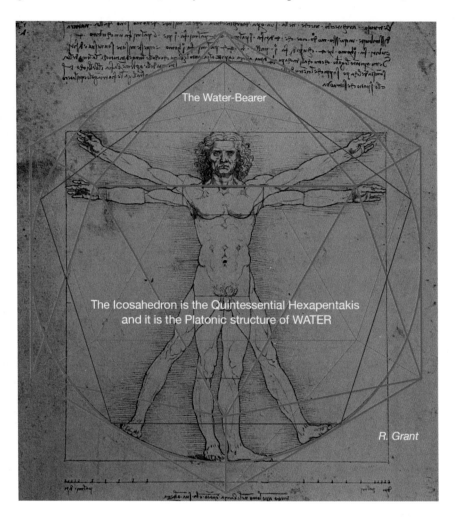

Figure 185: The different geometrical configurations, along with the figure itself, delineate an icosahedron solid.

The power of the icosahedron stems from its hexagonal-pentagonal configuration. It is made of 20 triangles. Every five triangles create a pentagonal pyramid, as shown below to the right. However, from a different perspective, its outline is clearly hexagonal, as shown below, to the left. It is a 3-dimensional version of the hexapentakis.

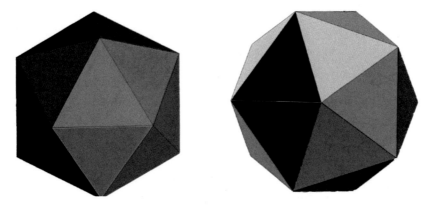

Figure 186: The icosahedron is a unique shape as it embodies the hexagon and the pentagon in its configuration. It is a 3- dimensional hexapentakis.

The hexapentakis is the heart and mind, coming together and unifying the two sides of the brain, bringing art, music, and emotion to harmony with science, math, and logic. It is the philosopher stone of the mind, a monument when the macrocosm and microcosm are united and a holistic view of the world is achieved, a moment of meaningful understanding.

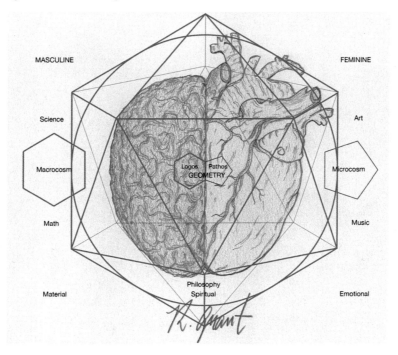

Figure 187: The hexapentakis represents the unification of the heart and the brain as well as the eyes of Ra and Thoth.

Starting with the square Leonardo drew in the Vitruvian Man, with sides of roughly 10 inches each, we can create a fractal of nested circles and squares embedded within each other. We can do the same thing, starting with the original circle of the sketch. These nested squares and circles create a fractal matrix with dimensions corresponding to many fundamental constants of nature.

As shown below, Davinci's circle (in green) has a circumference of 27.143, very close to $e/10$. The square, on the other hand, defines a circle that has a circumference of 31.41, which is also very close to $\pi/10$. As you may remember, the Euler number and π define the squareness and circularity through their relationship with the volumes of the cube and sphere. Uniting them together in this concealed fashion could very well be an attempt to square the circle, not through their areas nor their parameters, but through the fundamental constants they are defined with.

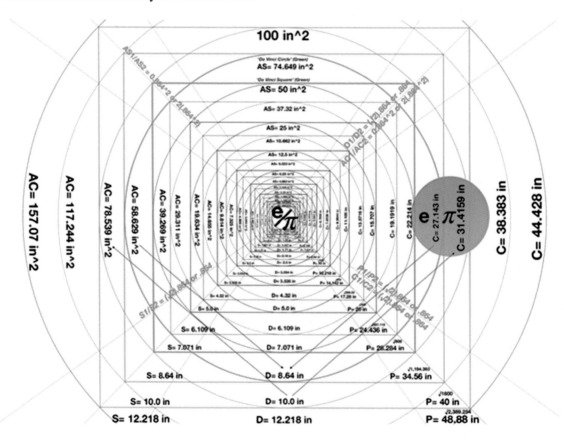

Figure 188: The square and the circle of Leonardo create fractals of squares and circles that define many mathematical and physical constants.

239

With this unique Squaring the Circle method, Leonardo, consciously or subconsciously, achieved the following:

- Mechanistically separated light and darkness, revealing knowledge of advanced mathematical/physics constants (including transcendental e, π, Φ, Ω, α (and many more.)

- Identified the mathematical connection between both transverse and longitudinal waves (scalar/square/cubic waveforms along with spiraling circular waves of light).

- Equilibrateed the universe to the light/darkness ratio of .864 to 1.00 and revealed holographic duality.

Through the Vitruvian Man, Leonardo conveys the divine balance of logos and pathos evolving into ethos. The conscious mind merges with the subconscious mind enabling the superconscious Heart-Mind experience. "the heart-brain." In short, Leonardo embodied through this masterful encryption in art precisely what Squaring Circle is intended to represent.

As it turned out, we can square the circle, just like Leonardo did, perfectly and using only a square and compass. The nine-step process requires the triangle, which turns it into an alchemical transmutation process, with the outcome being nothing but the philosopher's stone symbol. We start with two perpendicular lines, one horizontal, the horizon of the earthly realm, the other vertical, representing the heavenly realm. Together they define the point of origin, the center of perception.

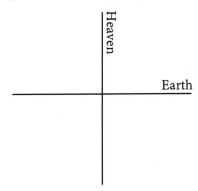

Figure 189: Laying out the heavenly and earthly domain via two perpendicular lines, intersecting at the point of origin.

From this center, we draw a circle of arbitrary units. Next, we enclose it with a square. In the fourth step, we draw the two diameters of the circle that run through the square's four corners, forming an X-shape. In the fifth step, we draw an isosceles triangle that passes through the intersection points between the diameters and the circle. From the corners of the base of the triangle, we draw two lines that are parallel to the original X-shape. In the seventh step, we draw two lines from the upper corners of the square such that they both meet at the center of the base of the triangle. We now connect the intersection points between the four new lines with the base of the triangle. Finally, in the ninth step, we complete the square, as shown below.

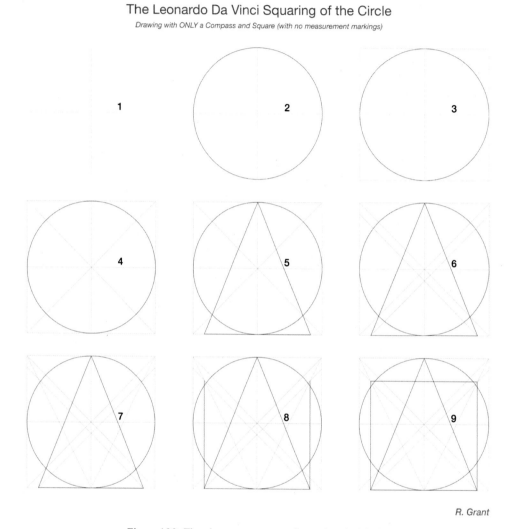

The Leonardo Da Vinci Squaring of the Circle
Drawing with ONLY a Compass and Square (with no measurement markings)

R. Grant

Figure 190: The nine-step process of squaring the circle .

When this configuration is overlaid on the Vitruvian Man, the two centers of the X-symbols coincide perfectly with the naval and groin of the man, which are also the two centers of the circle and the square, and with the two hands fitting perfectly between the lines of the two Xs, as shown below.

241

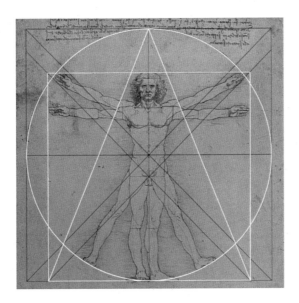

Figure 191: Overlaying the squaring of the circle symbol, the philosopher's stone, on top of the Vitruvian Man, reveals perfect matchings.

The triangle of the above symbol is identical to another one found in a drawing by the German alchemist Michael Maier, produced around 1618 (phi!). Interestingly, the title of the drawing is "The Production of the Philosopher's Stone." How appropriate!

Figure 192: "The Production of the Philosopher's Stone" by Michael Maier (1618).

And if the above was not enough, this same triangle is found on one of the most controversial objects in the world, the one-dollar bill, matching perfectly with its famous pyramid and the all-seeing eye, which also matches perfectly with the Vitruvian Man, as when the circle of Leonardo's sketch is made to have the same length of the pyramid, the square's lower side fits the base of the pyramid perfectly, with its higher side running right through the center of the eye, as shown below.

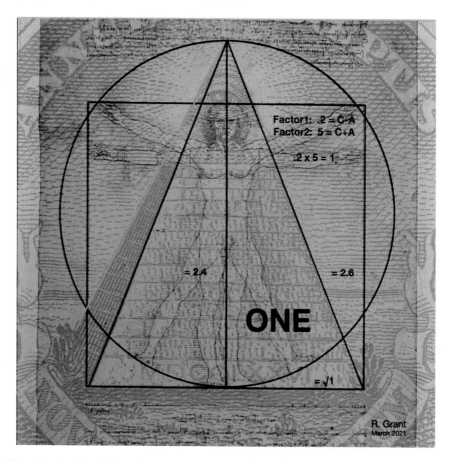

Figure 193: The triangle of the Philosopher's Stone matching perfectly with the pyramid of the one-dollar US bill.

Squaring the circle has always been considered an impossible task. Nevertheless, through Leonardo's sketch, we not only show that it can be done, but also that it can be done by hand, using a straight edge and a compass only. Moreover, finding the invisible triangle on the one-dollar bill, along with the square and circle, opens the door for so many speculations not only about the history of this bill but also about its connection with alchemy and Leonardo, to mention a few, as whenever a great discovery is made, many more mysteries emerge.

The Spiral of Theodorus

The spiral of Theodorus is a pattern made of adjacent right triangles, all having one side equal to number 1, such that the hypotonus of the previous triangle works as the side of the next, as shown below. The hypotenuses of the triangles are all square roots of numbers 2, 3, 4, …, etc.

Superimposing the Theodorus spiral on top of the Vitruvian Man reveals a perfect match between most of the spiral lines and the sketch, where the lines intersect with key points of the figure, such as the nose, fingers, feet, cut-lines intersections, etc., as shown below.

The head is exactly situated between the 24th and 26th lines of the spiral, forming a shape like a key-stone, centered exactly at the 25th line. Numbers 2 and 5 are mirror images of each other, whether in their shapes or in their reciprocals. It is as if Leonardo is saying one side of the brain is the mirror image or the reverse of the other.

Is this simply an unbelievable coincidence? Or did Leonardo really include all these elements, the hexapentakis, Great Pyramid, spiral of Theodorus, etc., in one simple sketch? Note this is a simple sketch on one paper in one of his many notes; it is not a full painting like the Mona Lisa or the Last Supper. He must have made the full design somewhere else and then copied to his notebook only those elements he wanted to become visible, such as the man himself and the square and the circle, being the keys to unlock all the hidden encryption.

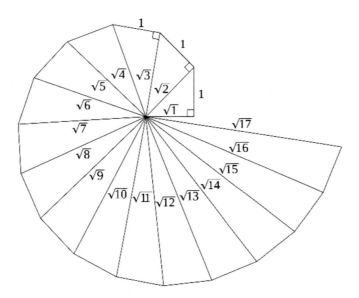

Figure 194: The spiral of Theodorus, made of adjacent right triangles having one side equal to 1 unit. In all the triangles, the hypotonus of the previous triangle works as a side for the next.

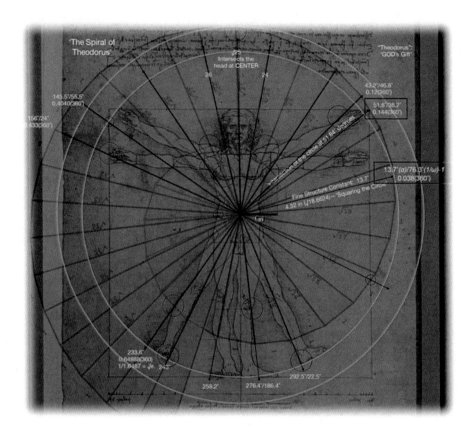

Figure 195: Superimposing the Spiral of Theodorus on the Vitruvian Man reveals perfect matching between most of the spiral arms and key points in the figure.

Leonardo's Cube of Delos

In the music part, we mentioned the story of the people of Delos who were asked by the priests at Delphi to double the cube of the temple of Apollo. The solution was to double the volume, which involved the cube root of number 2. It seems that Leonardo was also interested in this problem, as is seen from one of his sketches, shown below.

There are only three notes written on the page of the Vitruvian Man: a mysterious *1A* in the upper left-hand corner; a *1/14* (written in backward mirror text) along with a descriptive text outlining the *14* separations of man; and number *126* (not backward) in the upper right-hand corner of the sketch next to a clear bookmarking fold line. Of course, the answer to doubling the cube was to increase its length from unity to $\sqrt[3]{2}$ (the cube root of 2), equaling to 1.26, which is also the new Precise Temperament tuning.

245

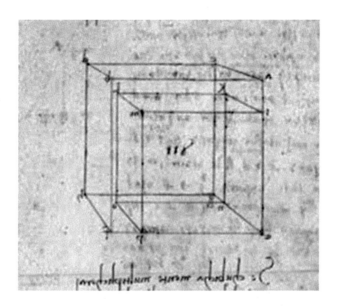

Figure 196: Leonardo's Cube of Delos sketch along with the number 126 at its center.

Could this be what da Vinci was trying to tell us after all, by placing a decoy page number next to the bookmarking fold line, possibly signifying an unfolding to a higher octave? It could very well be, even though we may never know for sure.

Leonardo's Flower of Life: The Singularity Convergence

As we saw earlier, the Flower of Life is one of the most sacred symbols of antiquity. We elaborated further on its uniqueness by showing how it can be used to solve for prime factorization, especially through the right triangle relationship.

Leonardo's notebooks include many pages that exhibit different forms of this symbol, indicating how much it resonated with him. One of these sketches, shown below, is of particular interest. Notice the intersections of the lines for the equilateral triangle (which is also two adjacent right triangles). Notice also the letters c, d, and e (backward and upside-down), the numeric values of these numbers are [3, 4, 5], the same numbers of the first Pythagorean triangle (a coincidence?).

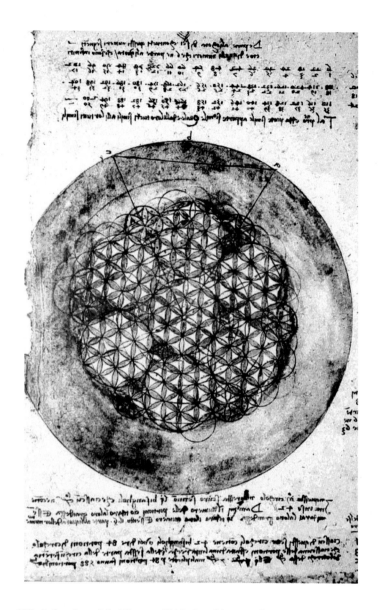

Figure 197: A drawing of the Flower of Life found in one of Leonardo's notebooks.

Da Vinci's Flower of Life is deeply related to the geometries associated with the base 12 numerical system, along with connections to musical notes, light (reflected and absorbed), as well the zodiac, months of the year, and even days of the week. Most importantly is the new chordal relationship between numbers, mainly between the triplet group of [1, 4, 7], [2, 5, 8], and [3, 6, 9] and the new numbers of phio, sieve, and eno, positioned between 3 and 4, and 6 and 7, and after 9. This also relates to the decimal-based metrology versus the imperial system. The base-12 system is embedded within the base-10 just as there are 432° positions embedded within the 360°, which is probably also why we have 1,296,000 seconds in a compass, breaking evenly into 432,000 seconds, 864,000 seconds, and 1,296,000 seconds to mark an equilateral triangle.

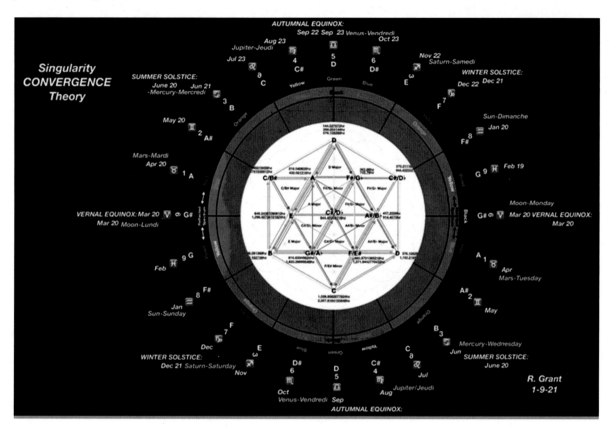

Figure 198: Superimposing the various disciplines of science we covered in this book on top of Leonardo's Flower of Life, with all converging into one beautiful image of unity

These correspondences also inform the differentiation characteristics of electric force, the only force of the universe, which differentiates into electricity/magnetism, radiation (light/no sound/transverse waves), and gravity (dark/sound compression/longitudinal waves), as we discussed earlier. It also creates matter and spin (electron, proton, and neutron, and their mirrored opposites: anti-particles).

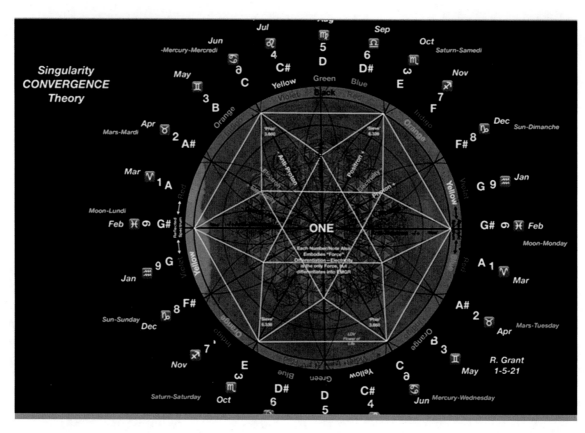

Figure 199: The Flower of Life symbol can be used as a blueprint for the various elements and forces of physical reality, such as the elementary particles and the forces of electromagnetism and gravity.

Whether all the above were intended by Leonardo or not is not what really matters. Even if very few of the above were originally in Leonardo's head when he drew the Vitruvian Man or the Flower of Life, still the fact that the main elements of every scientific discipline, including mathematics, geometry, music, or physics, can be superimposed on top of one or two images is a strong indicator of their unity, as this convergence of ideas and principles could only be achieved if they were to emanate from a single source. And these types of images drawn by visionary individuals, as well as the countless ancient buildings and edifices, are nothing but encryption keys that, once studied with due respect and appreciation, will help us decipher the message, which, in turn, will lead us toward the original source of this knowledge and wisdom, to our ultimate destination, that of the divine itself.

Expanding Consciousness

Hand-Drawn Geometrical Patterns by

Robert E. Grant

"Drawing is putting a line around an idea."

-Henri Matisse

One decisive factor in becoming a polymath, having a broad understanding of many disciplines, is balancing the two sides of the brain. It is believed that the left side of our brains is reserved for all tasks that are devoid of emotions, like math, intellectual thinking, rational decisions, etc. The right side, on the other hand, is reserved for those tasks that require feelings and artistic skills. Therefore, to have a balanced brain, we need to work on both sides at the same time, be rational and emotional, scientific and artistic.

One perfect way to activate the emotional side is to do art, whether drawing, sculpting, writing, playing music, etc. Most visionary thinkers of the past had excellent or outstanding artistic skills. Many were also musicians or at least played music.

Drawing is probably the king of these right-brain skills. It stimulates imagination and creativity, enhances eye-hand coordination, and evokes feelings and emotions. Using colors is like playing music; for each color, there are many tones and undertones, etc. And just like there are many branches of science, there are also many ways and fields of drawing and painting. Both writers of this book are accomplished artists, whether in the art of comics creation, as is the case of Talal Ghannam, or in the art of geometrical drawing, as it is for Robert Grant.

Becoming a comic creator requires combining many artistic skills. One needs to be a good artist to create interesting characters. He needs to understand emotions and how to convey them using facial and bodily expressions. He will have to develop spatial perspectives for setting the stages, directing the scenes, etc.

250

Drawing animal and human figures require some formality about their anatomy as well. And when the artist is also the author, this entails good writing skills as well as broad imagination. All of this will ensure the right side of the brain is fully occupied and functional.

Using the compass and straight edge to draw geometrical shapes is by itself a challenge that requires modifying one's perception to be able to see through all the lines and points of intersection to filter out shapes that are not by any means obvious. This skill is often rewarded by discovering interesting geometrical forms that expand the consciousness and enable the discovery of novel forms of higher dimensionality. It also entails discovering novel mathematical solutions to problems that are thought to be very difficult or even have none.

Below we present some of the art made by Robert E. Grant, especially those falling in the context of this book. Through these images, we can get a sense of elegance, as well as dimensionality, where much of the images seem to pop out of the page as if they acquire a third dimension that only activates through gazing. Even though most of these figures initiate from the same hexagonal, octagonal, or higher matrix, the progress always leads to a new shape that most of the time has never been recognized before. And while it is not the intention of the artist, however, each of these drawings hides a key that awaits to be discovered. It is the duty of the gazer to look beyond the obvious and through the simple lines to uncover the hidden, hold on to it, and unravel its secrets.

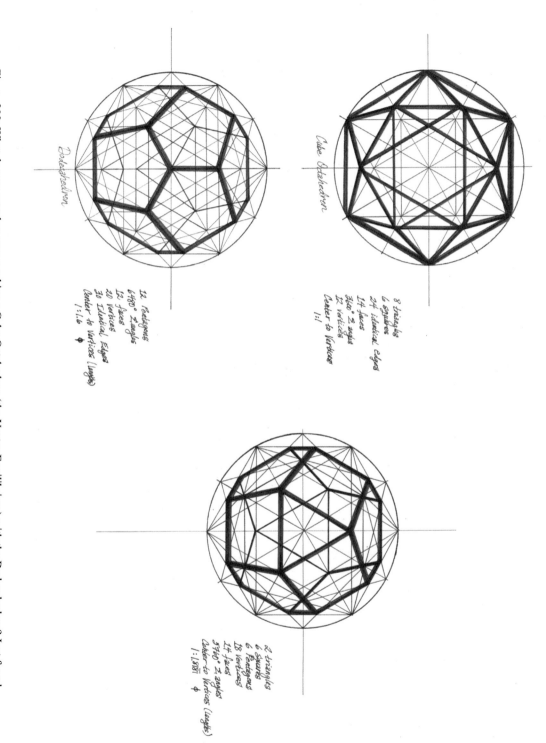

Figure 200: What do you get when you combine a Cube Octahedron (the Vector Equilibrium) with the Dodecahedron? Just found this Johnson Solid (pictured on the right: 'Gyrobicupola') inside the circle intersections of the Flower of Life. This geometry appears to possess very unique properties related to 24-ness, and its interior angles also sum to 5,760°.

Cube Octahedron

8 triangles
6 squares
24 identical edges
14 faces
3120° ∠ angles
12 vertices
Center to Vertices
1:1

Dodecahedron

12 Pentagons
6480° ∠ angles
12 faces
20 Vertices
30 Identical Edges
Center to Vertices (Lengths)
1:1.6   φ

2 triangles
6 Squares
6 Pentagons
14 faces
18 Vertices
5760° ∠ angles
Center to Vertices (Lengths)
1:1.818̅   φ

252

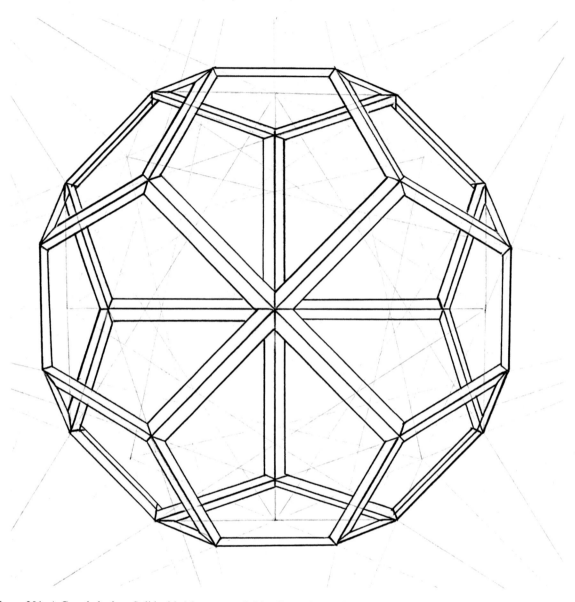

Figure 201: A Granthahedron Solid with 16 pentagonal sides (8 regular (5 sides of 1 unit each), and 8 irregular (2 sides each of 1.718 (Euler-1)). Sum of Interior Angles: 8,640°.

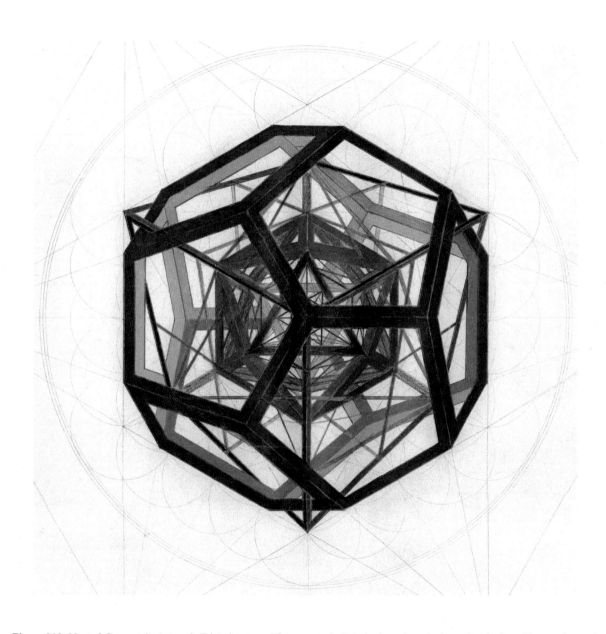

Figure 202: Nested Geometric Artwork (Listed outward from center): Tetrahedron, Icosahedron, Octahedron, Hypercube, and Dodecahedron.

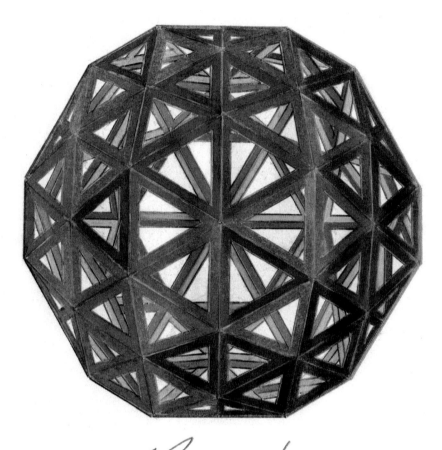

R. grant

Figure 203: The Truncated Icosahedron (Hexapentakis).

Figure 204: Framework design of a Water Wheel with 24 stellations around a central icosahedron.

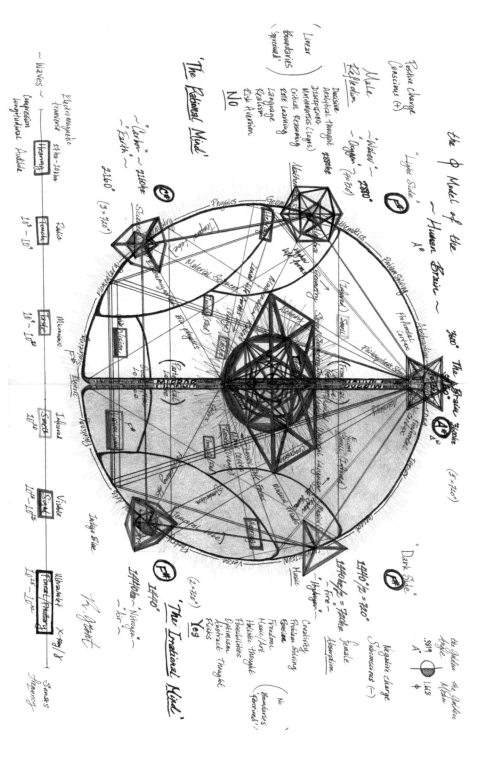

Figure 205: Is the brain a physical representation of the Greek letter for the Golden Number Φ (a circle with a vertical line running through the center)? Balance of the mind centers across the left brain (rational thought, logic, mathematics, and sciences-conscious thought), and right (creative expression, intuition, art, and music-subconscious expression) activates the neural net within the Corpus Callosum (a broad band of nerve fibers joining the two hemispheres of the brain), Pineal and pituitary glands, allowing each to fully function, operate and receive. This balance also allows for neural pathways to extend to the Prefrontal Cortex, where higher-order thought and philosophy (the Love of Wisdom) develops. Just as exercising one side of our bodies only leads to atrophy and imbalance on the other side of the body and poor athletic performance as a result, the mental athlete knows that rhythmic balanced interchange within the brain receiver is the key to mastery. Basic and balanced understanding in both the sciences and arts is critical and often overlooked in today's reductionistic and highly specialized university curricula. So it is with thought and universal connection. Universal Beauty is the perfect balance of Art within Science and Science within Art.

Figure 206: Complex nested geometric structure (outward from center): Star Tetrahedron, Cuboctahedron, Icosahedron, Hypercube, Octochoron Polytope, and Truncated Icosahedron (hexapentakis).

Figure 207: The Flower of Life informs all regular polyhedra geometric forms from its wave intersections: (Outward from center): Tetrahedron, Octahedron, Icosahedron, Cuboctahedron, and Dodecahedron.

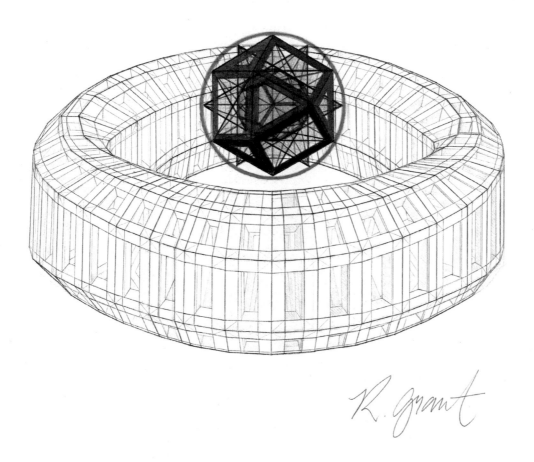

Figure 208: The Torus in a stylized architectural illustration with a levitating sphere containing the cuboctahedron, the Star Tetrahedron, and the Dodecahedron.

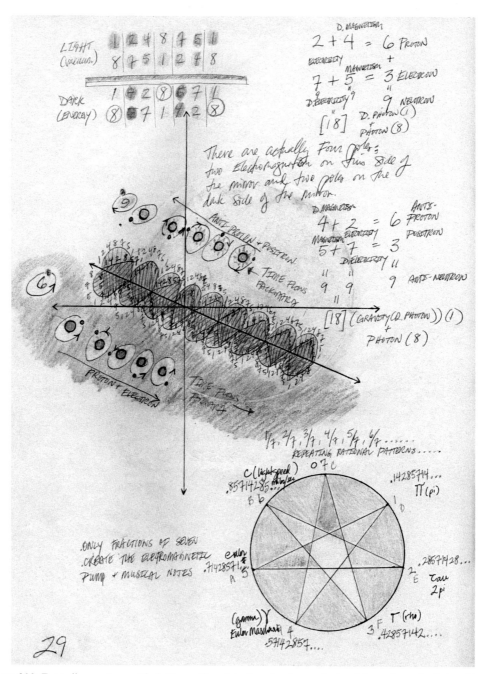

Figure 209: Extending on James Clerk Maxwell's original quaternion analysis of the reciprocal relationship of radiation and gravity.

261

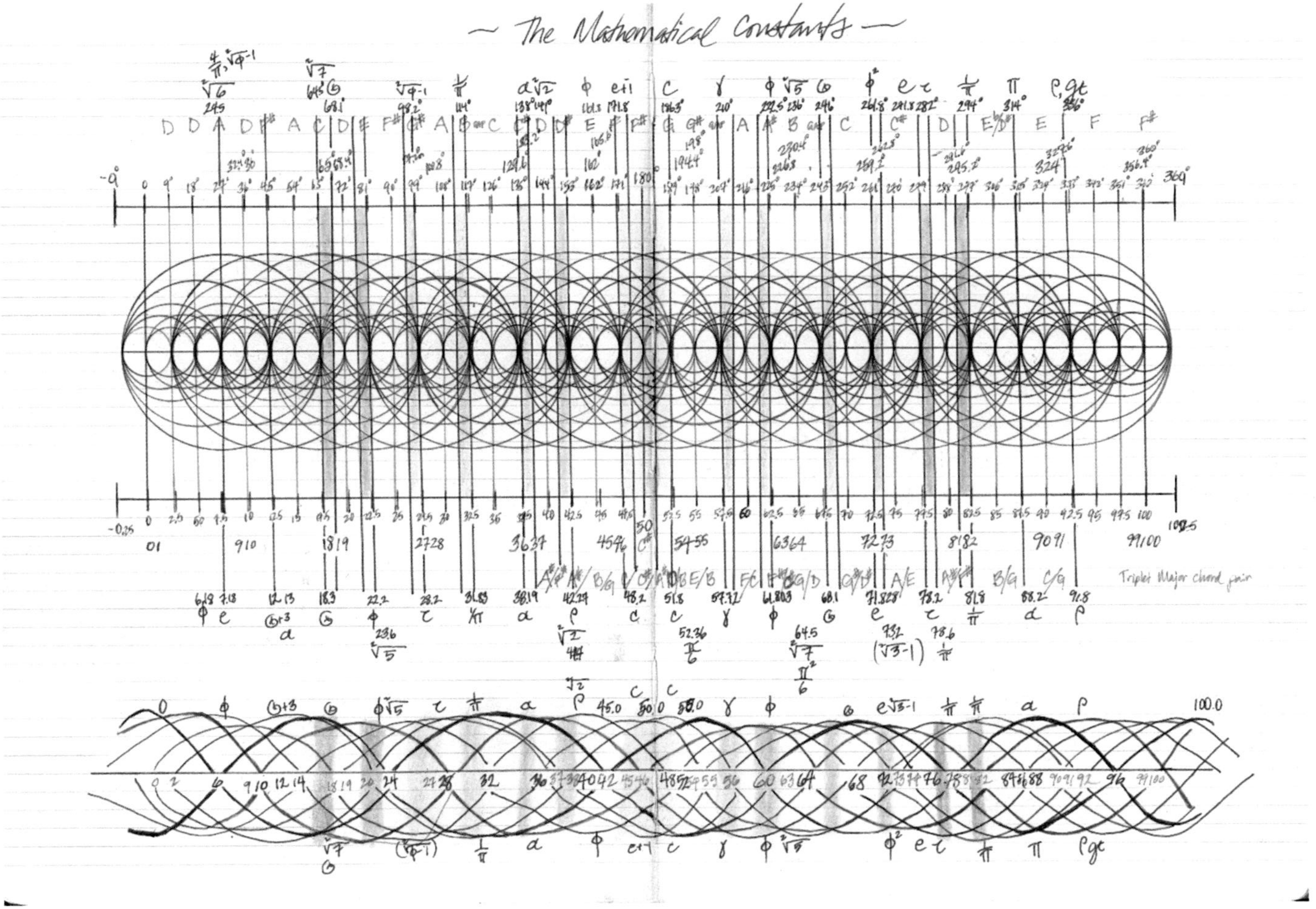

Figure 210: Mathematical constants are dimensionless and reflect each other in a wave-based four-fold reciprocal mirror symmetry.

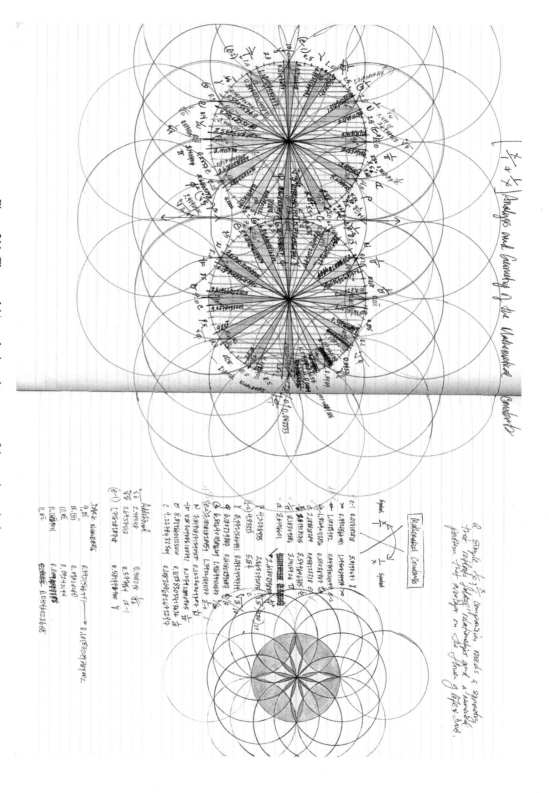

Figure 211: The x and 1/x analysis and geometry of the mathematical constants.

263

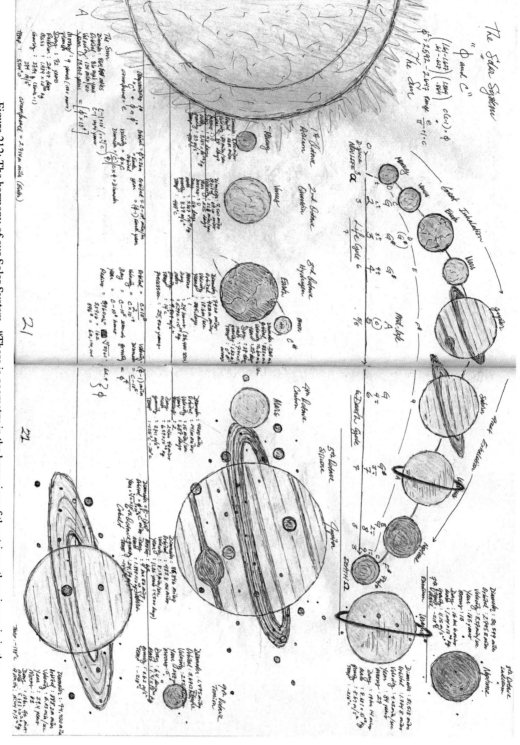

Figure 212: The harmony of our Solar System. "There is geometry in the humming of the strings, there is music in the spacing of the spheres." - Pythagoras.

264

Figure 213: From the balancing of masculine and feminine energies emerges the final Tau.

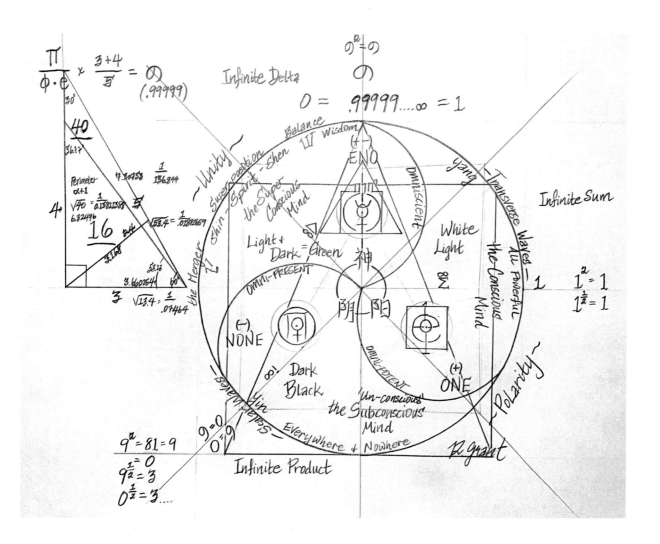

Figure 214: Transition from polarity to unity consciousness.

266

Figure 215: The Archetype of the Higher Self

THE SPIRAL OF CONVERGENCE

ALL IS ONE

The flow of the ideas and discoveries presented in this book appears to converge toward a *singularity* point where math, geometry, physics, alchemy, chemistry, and sound become indistinguishable from each other, albeit perceived from different perspectives (by each observer) as fractional divisions of the number 1. The agents or mediums of this unity are numbers, whether odd, even, rational, irrational, etc., along with geometry, especially circles and triangles.

The mathematical and physical constants, in particular, are one critical element in this cycle of convergence, emanating from each other through simple algebraic operations as if via numerical alchemical transmutation. One such perfect cycle starts from the one (the number 1). From it comes the tetractys, the triangular configuration of numbers 1, 2, 3, and 4, adding up to 10 (which is simply 1 in the digital root sense). Their multiplication produces 24, the important number that defines prime numbers' geometry, the Fibonacci sequence, and the Flower of Life, to mention a few.

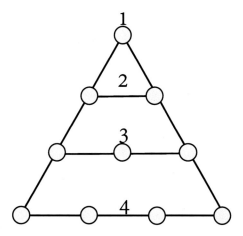

Figure 216: The tetractys: a triangulation of dots that represent numbers from 1 to 4.

Number 24 has many unique characteristics; of these is its reciprocal: $1/24 = 0.04166...$, which can be approximated to 0.042, identical to its mirror image. The ratio of 24 to 41.666... is 5.76... very close to the Euler-Mascheroni constant, which is also almost identical to 24×0.024. The exact value of the Euler-

Mascheroni constant is 0.5772…, and if we divide it by 42, we get 0.01374…, a fractal of the golden angle (137.5) and one manifestation of the golden section. This particular cycle of transmutation from number 1 to 24 to Euler-Mascheroni to 137.5, and finally Φ is another example of the convergence theory, which we touched upon in the wave theory of constants, stating that the most fundamental constants are but simple transformations of each other, especially the golden section phi, the most probable origin of all other constants.

The best geometrical representation of this convergence would be through the form of a spiral, the most natural and observed form in nature, existing on all levels and transcending scientific boundaries, from quantum mechanics to general relativity, from the microscopic to the macroscopic. This spiral stretches outwardly in a very careful and calculated twist that ensures the many is still bounded toward the one, the singularity point; not too close to collapse back into the singularity, nor too far to diverge into unbounded infinity.

The stretching and twisting are controlled by the two main constants of nature, the circular π, and the linear e; the former ensures the spiral is twisting in a circular motion, providing the freedom that allows creativity and individuality, while the latter makes sure that this freedom remains bounded as not to lose its path away from the origin. Within its twisting arms, all knowledge is coded in space and time, hidden beneath layers of numbers and geometry.

But how do we start constructing such a spiral? From prime numbers, perhaps? Or maybe from the right triangles that define them? What proportionalities would such a fundamental triangle have?

Maybe we should start from the most basic form of an equilateral triangle, the one used in forming the tetrahedron and octahedron of the Platonic solids, the same triangle emanating from 1, the tetractys. Halving it creates the right triangle shown below with its dimensions chosen such that the base of the triangle will have the rational value of unity. This translates into a value of 2 for the hypotenuse, leaving the fundamental $\sqrt{3}$ for the height.

We can transform this fundamental equilateral triangle into a spiral using the same logic implemented in creating the spirals of Theodorus and Regulus. As shown below, we start from the equilateral triangle, stacking bigger versions of it, one next to the other, such that the side of the previous triangle will work as the height of the next, as shown below. Note that the ratio between the height of the triangle and its base, for all triangles, is always $\sqrt{3}/2$. Thus, if we start from a triangle of $[1, \sqrt{3}, 2]$ units, the next one will have a hypotenuse of $4/\sqrt{3}$, and so on.

Notice how the saw-like lines delineate a smooth spiral inside them, with the 12th triangle coming back perfectly horizontal, matching up with the height of the smallest initial triangle. This is another geometrical representation of the 12-based numeric system.

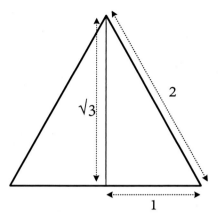

Figure 217: Starting from an equilateral triangle with sides' of 2 units in length, we create two right triangles of dimensions [1, √3, 2].

This perfect spiral is logarithmic in nature because its dimensions satisfy the relationship $r = e^{(\alpha+\theta)}$, where r is the radius from the center of the spiral to any points on its arms, and α is its rate of increase. The θ variable is the angle, in radian, at which the spiral turns, which for our case is equal to $\pi/6$ (30°). Therefore, we have both our fundamental constants e and π ruling the shape of this spiral and controlling its expansion, as we postulated they should.

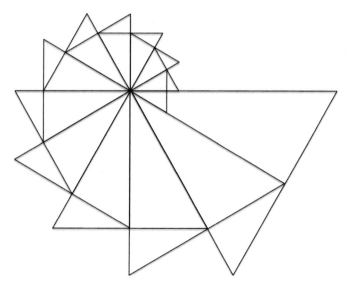

Figure 218: Using equilateral triangles as a grid to create a perfect logarithmic spiral, guided by their halves, the right triangles.

The number *e* can also be found within the dimensions of rectangles that enclose the spiral. From the image shown below, we see how a rectangle that touches the curvature of the spiral at its four sides has dimensions of $1/(\sqrt{e} - 1)$ and $\sqrt{e} - 1$. Notice that $(\sqrt{e} - 1)^2$ is simply 0.42.

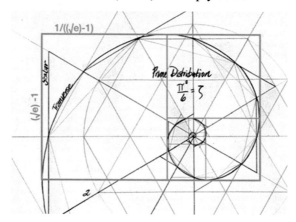

Figure 219: The Euler number, emerging from the dimensions of rectangles touching the curvature on all sides.

The golden section is embedded within this logarithmic spiral as well, emerging from those rectangles having two parallel sides touching the curvatures of the spiral, with the other two sides matching the bases of the triangles, as shown below to the right. Note that these rectangles are not golden spirals, as the two sides' ratio is not Φ but Φ^2. Thus, phi emerges from combining the linear and circular aspects of the spiral, from the merger of *e* and π.

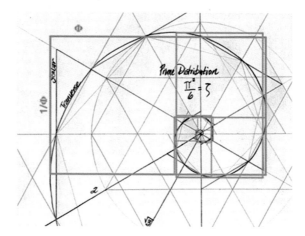

Figure 220: Drawing rectangles such that two sides touch the spiral and the other two sides match the bases of the triangles results in dimensions that exhibit the golden section.

Even the new numbers of phio and sieve can be found embedded within the turns and twists of this amazing spiral through the ratios of a rectangle whose one side touches the curvature of the spiral and the other three matching the bases of the triangles, as shown below.

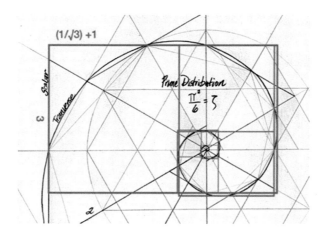

Figure 221: The new numbers of phio and sieve emerging from the dimensions of rectangles having three sides matching the bases of the triangles and one side touching the curvature of the spiral.

Other fundamental numbers can be found as well. For example, if we start with a unit value of 1 in the spiral, the 180° opposite length is 2.37037, which is 1+(α/100), with α being the fine structure constant 137. Its perpendicular value (preceding clockwise rotation pointing north) is the square root of that value, which is precisely equal to Lieb's Square Ice Constant: 1.539600...

The value of $\sqrt{(4/3)} = 1.1547...$ is the ratio of the base to the height of the triangles, and it governs how each new triangle grows from the previous one. It acts like another Euler number in logarithmic expansions and accounts for each expansion precisely at 30° arc intervals around a circle. The natural log can land at the same locations as the log based on ($\sqrt{(4/3)}$), but uses base 12 expansion values to achieve the same. For example, $e^{0.4315} = [\sqrt{(4/3)}]^3$, both landing at 1.539600... (which is also Lieb's square ice constant to infinite accuracy).

Notice how 0.4315... is very close to the fundamental 432 Hz of Pythagorean tuning. And as we showed earlier, 432 emanates from the 12-based system as 432/1.2 = 360, which is the angular value of the 10-based system. Now, if we subtract 1 from the value of 1.5396007..., we get remarkably close to the Planck time constant, being the amount of time required for light's wave perturbation to traverse one Planck length (0.539×10^{-43} seconds), which is even more interesting when we consider that the corresponding horizontal axis of the spiral is 2.37037, being the inverse of the fine structure constant that defines the separation of light from the darkness. So, the vertical axis shows a relationship to time, and

272

therefore scalar longitudinal/compression, and the horizontal axis is a transverse wave and a direct relationship to light.

Thus, this spiral combines scalar and transverse wave functions into one field, which are inherent to the structure of the right triangles that delineate the smooth circular curvature of the spiral. As we explained earlier, scalar waves are compression/longitudinal waves of gravitation (phonon-sound waves which carry mass), and transverse waves are spiral electromagnetic waves (radiative photons of light), in which transverse wave (excitation) perturbations are precisely perpendicular to the gravitational scalar waves. The scalar structure of spacetime surrounds and interconnects all the transverse waves of light and matter we perceive in the material world around us, which we call the Universe. The scalar waves are always there as set mathematical intervals that are fixed within the spacetime structure. Only certain ratios can be achieved with right triangles and whole numbers. The scalar, therefore, defines the transverse and vice versa.

It is amazing how all of the above, and more, emerge from one simple spiral generated from a simple triangle, all emanating from a single point, the singularity point from which all numbers and knowledge diverge.

In mathematics, the term *singularity* refers to certain values for which a function would become undefined or unbounded, diverging toward infinity. Recently, the same term has been coined to describe the hypothetical moment when artificial intelligence (AI) would reach the point of self-awareness. Our *convergence singularity* is not that far from both meanings, as it defines the moment when humanity becomes self-aware of its potentials as well as its position in the universe, when its understanding and knowledge will transcend the dominant materialistic perspective and diverge into higher dimensions, ushering a transition from darkness to light.

The ancient notion of light emerging from darkness seems to be a very prolific linguistic theme across many languages, cultures, and religions. For example, the Sanskrit word *Gu-ru* means to bring light (Ru) from darkness (Gu). From AχΩ to Yin-Shen-Yang to Mer-Ka-Ba, the deeper symbology seems to be amazingly consistent. Each appears to have a dualistic binary construct that evolves into a trinary relationship between syllables. In the case of Yin-Yang, its binary iteration includes the Chinese characters for both the sun (Yang) and the moon (Yin). However, when Shen arrives, it complements and evolves the word combination to a trinary relationship, with Shen introducing the Chinese characters for both spirit and earth (heavenly earth). Furthermore, the Ka in the ancient word Merkaba connotes the bringing of the spirit into a union with light (Mer) and with the body of flesh (Ba). It is the Φ coming to fulfill the π-e relationship ($\pi\Phi e$). (Notice the circular and linear aspects of the phi letter.)

In Plato's *Timaeus*, it is explained that the two bands that form "the soul of the world" cross each other like the letter χ (chi). Plato's analogy brings an expanded view to the early Christian meaning of AχΩ. In the Egyptian pantheon, the sun god Ra is merged with the left eye of Horus, creating Ra-Horakhty, a deity often symbolized by the flying sun disk and possessing powers of the two prior divine emanations. It is quite astounding that the respective mythologies (of very disparate lands and teachings), their various spiritual symbols, and even the differing (or very similar) phonetics all seem to possess similar root cognates, characteristics, and philosophies, and importantly, also seem to be pointing to a convergence of unity and balance that goes beyond and transcends the very concept of duality itself.

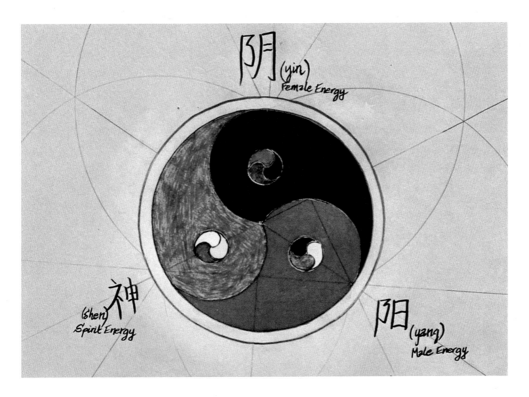

Figure 222: The Shen symbol representing the unity of the trinary aspects of the sun, moon, and earth.

In terms of simple numerals, the above trinary principle will correspond to numbers [0, 0.999..., 1]. Numbers 1 and 0 are the polar aspects of our modern era. They are the binary representation of the digital age we are all living and experiencing to its maximum, where the internet and social platforms have transformed our lives through the information revolution. Unfortunately, though, this binary approach has produced many negative aspects alongside the positive ones. By dint of the above, a third element is needed to restore the balance. The eno number of (0.9999...) corresponds to this third element, where it stands for the hidden aspects of the soul or spirit that we need to bring back into the abstract and mundane digital age for our lives to have a greater meaning again. It is the number between 1 and 0 that combines both aspects, 1 in its approximated value and 0 in its digital root; it is the superposition value of the quantum computing age of 1, 0, and x.

But should this quantum revolution bring light with it or more darkness?

It is up to us to decide, as has always been the case. Quantum computing will definitely bring so much advancement to all fields of science; however, it will also bring lots of risks, which include rendering much of current encryption systems vulnerable if not useless. This will either usher in an age of chaos or an age of better encryption as well as better privacy and sovereignty over our own personal data, especially our DNA. The convergence methodology is definitely the right step toward such sovereignty. It is an already overdue new renaissance that should *spiral* humanity *into* awareness and control and toward a better future for the whole earth. This is achieved by keeping the spiral tethered to its origin, to Φ, bringing both π and e, left and right, advancements and morality, into a perfect balance, which was the intention of this book from the beginning. And while we do not claim we covered everything in the book, nor that everything we covered is 100% true, we are confident we made the right step, hoping it was at the right moment, which only time will tell.

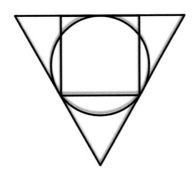

... So Below.

Bibliography

Adair, Robert K. *The Great Design: Particles, Fields, and Creation.* Oxford University Press, USA, 1989.

Baigent, Michael and Richard Leigh. *The Elixir and the Stone: The Tradition of Magic and Alchemy.* Arrow, 2005.

Bartholomew, Alick. *Hidden Nature: The Startling Insights Of Victor Schauberger.* Floris Books, 2003.

Branderburg, John. *Beyond Einstein's Unified Field: Gravity and Electro-Magnetism Redefined.* Adventures Unlimited Press, 2011.

Coates, Austin. *Numerology.* Carol Publishing Corporation, 1991.

Cook, Theodore Andrea. *The Curves of Life: an Account of Spiral Formations and Their Application to Growth.* Constable and Company, 1914.

Dunlap, R. A. *The Golden Ratio and Fibonacci Numbers.* World Scientific Publishing Company, 1998.

Du Sautoy, Marcus. *Symmetry: A Journey into the Patterns of Nature.* Harper Perennial, 2009.

Glynne-Jones, Tim. *The Book of Numbers.* Arcturus Publishing Limited, 2007.

Hancock, Graham. *Fingerprints of the Gods.* Three Rivers Press, 1996.

Hayes, Michael. *The Hermetic Code in DNA: The Sacred Principles in the Ordering of the Universe.* Inner Traditions, 2007.

Ghannam, Talal. *Geonumeronomy.* KDP Publishing, 2021.

Hayes, Michael. *The Infinite Harmony: Musical Structures in Science and Theology.* Weidenfeld & Nicolson Ltd, 1995.

Heath, Richard. *Sacred Number and the Origins of Civilization: The Unfolding of History through the Mystery of Number.* Inner Traditions, 2007.

Jenny, Hans. *Cymatics: A Study of Wave Phenomena & Vibration.* Macromedia Press, 2001

Krafft, Carl Fredrick. *The Ether and its Vortices.* Borderland Sciences, 1987.

Kramer, Samuel Noah. *History Begins at Sumer: Thirty-Nine Firsts in Recorded History.* University of Pennsylvania Press, 1988.

Lundy, Miranda. *Sacred Number: The Secret Quality of Quantities.* Walker & Company, 2005.

McTaggart, Lynne. *The Field: The Quest For The Secret Force Of The Universe.* Harper Perennial, 2002.

Milton, Richard. *Alternative Science: Challenging the Myths of the Scientific Establishment.* Park Street Press, 1996.

Milton, Richard. *Shattering the Myths of Darwinism.* Park Street Press, 2000.

Mitchell, John and Brown, Allan. *How the World Is Made: The Story of Creation According To Sacred Geometry.* Inner Traditions, 2009.

Olsen, Scott. *The Golden Section: Nature's Greatest Secret.* Walker & company, 2006.

Shesso, Renna. *Math for Mystics: From the Fibonacci sequence to Luna's Labyrinth to the Golden Section and Other Secrets of Sacred Geometry.* Weiser Books, 2007.

Skinner, Stephen. *Sacred Geometry: Deciphering the Code.* Sterling, 2009.

Stewart, Ian. *Why Beauty Is Truth: A History of Symmetry.* Basic Books, 2008.

Sscwaller de Lubicz, Rene A. *A Study of Numbers: A Guide to the Constant Creation of the Universe.* Inner Traditions International, 1986.

Tweed, Matt. *Essential Elements: Atoms, Quarks, and the Periodic Table.* Walker & Company, 2003.

Wells, David. *Prime Numbers: The Most Mysterious Figures in Math.* Wiley, 2005.

Plichta, Peter. *God's Secret Formula: The Deciphering of the Riddle of the Universe and the Prime Number Code.* Element Books Ltd, 1998.

INDEX

287